The MACAT Library
世界思想宝库钥匙丛书

解析托马斯·库恩
《科学革命的结构》

AN ANALYSIS OF
THOMAS KUHN'S
THE STRUCTURE OF
SCIENTIFIC REVOLUTIONS

Jo Hedesan　Joseph Tendler ◎ 著
丰国欣 ◎ 译

目 录

引言	1
托马斯·库恩其人	2
《科学革命的结构》的主要内容	3
《科学革命的结构》的学术价值	5
第一部分：学术渊源	7
1. 作者生平与历史背景	8
2. 学术背景	13
3. 主导命题	18
4. 作者贡献	24
第二部分：学术思想	31
5. 思想主脉	32
6. 思想支脉	40
7. 历史成就	48
8. 著作地位	54
第三部分：学术影响	61
9. 最初反响	62
10. 后续争议	70
11. 当代印迹	76
12. 未来展望	81
术语表	85
人名表	92

CONTENTS

WAYS IN TO THE TEXT 101
    Who Was Thomas Kuhn? 102
    What Does *The Structure of Scientific Revolutions* Say? 103
    Why Does *The Structure of Scientific Revolutions* Matter? ... 106

SECTION 1: INFLUENCES 109
    Module 1: The Author and the Historical Context 110
    Module 2: Academic Context 116
    Module 3: The Problem 122
    Module 4: The Author's Contribution 129

SECTION 2: IDEAS 135
    Module 5: Main Ideas 136
    Module 6: Secondary Ideas 145
    Module 7: Achievement 154
    Module 8: Place in the Author's Work 161

SECTION 3: IMPACT 169
    Module 9: The First Responses 170
    Module 10: The Evolving Debate 179
    Module 11: Impact and Influence Today 186
    Module 12: Where Next? 192

Glossary of Terms 197
People Mentioned in the Text 205
Works Cited 213

引 言

要　点

- 托马斯·塞缪尔·库恩（1922—1996），美国物理学家、历史学家和科学哲学家。
- 在《科学革命的结构》（1962）中，他认为科学知识的所有进步都源于科学知识的"革命"。
- 此书改变了科学的观念，同时也改变了深受科学家及其方法影响的科学的实践方式。

托马斯·库恩其人

托马斯·库恩于1922年出生在俄亥俄州辛辛那提市。他的父亲塞缪尔·L.库恩是一位工业工程师，母亲名为米内特·斯特洛克·库恩；他后来把母亲描绘为"家庭知识分子"[1]。1962年，库恩出版了《科学革命的结构》，时年40岁。

库恩高中毕业时就立志探索数学和物理。他在美国东海岸上完寄宿学校后，于1940年上了哈佛大学*。他获得本科学位后在军队服役两年，从事雷达技术工作。

1945年第二次世界大战*末期，美国决定部署原子弹摧毁日本的两座城市，这提醒了全世界，科学的力量影响人类的命运，其责任重大。库恩决定涉猎这一问题，他不再研究科学的运用，而去探索科学背后所隐含的法则。

战争结束后，库恩重新开始学业，并于1949年获得哈佛大学博士学位，其导师是时任哈佛大学校长的科学家詹姆斯·布赖恩特·科南特*。这位曾参与原子弹研发的科学家招募了许多知名学者，试图努力普及科学，当时的托马斯·库恩虽然寂寂无闻，也在其中。

由于在很大程度上受到科南特的影响，库恩把教授科学史和科学哲学*作为自己的专业领域。在库恩出版《科学革命的结构》之前，科学史*被认为是一项副业——虽然有趣，但不值得发展成一门学科。库恩的工作改变了这一认识，并将很多科学概念引入主流思想，包括"范式"*（大致指一种解释科学实验结果的知识模型）、"科学革命"*（一种范式取代另一种范式的时刻）和"范式转移"*（一场科学革命后的智力和科学结果）等等。

《科学革命的结构》的主要内容

科学家一向把自己的工作看作是线性的，每一次新的进步都会增加并优化自古以来所积累的科学知识储备。然而，在《科学革命的结构》中，托马斯·库恩却认为科学是周期性的，也就是说，在各个重复阶段里都会产生进步。

如库恩所言，一旦科学家推翻现行范式，"常规科学"*时期就结束了，代之而起的是"革命科学"*时期。革命时期的突破并不是建立在已有知识储备的基础上，并永久地改变了科学家理解世界的方式。除了"范式"和"革命科学"这两个概念以外，库恩还采用了第三个概念："不可通约性"*。

科学家如果想推翻现行范式就必须取代它，但是新范式往往对现实作不同或者相反的解释，因而是"不可通约的"，即新旧两种范式没有以共同或共享的方式来解释现实。所以，在科学革命末期，选择"共识范式"*，既是科学家的选择问题，也是基于范式力量解释现实的问题。

库恩断言，如果任何特定时代的科学家用相同的方式看待世界，那是因为他们使用了相同的范式。但这并不意味着，客观世界和他们所观察到的一模一样，世界的现实是独立于观察者而存在的。范式一旦发生变化，改变的不是世界本身，而是科学家，科学家从一个不同的视角观察世界。这是库恩的读者常常没有理解的一点，他们以为库恩所说的"范式转移"指的是世界变化事件。

库恩用历史来证实自己的理论。18世纪，法国化学家安托万·拉瓦锡*发现了氧。当然，在这次发现之前，氧早已存在；在拉瓦锡出生之前，人类就一直呼吸氧气。库恩解释说，拉瓦锡的发现改变了范式。世界没有变化，但科学家们已不再用以前的方式审视它了。

《科学革命的结构》自从初次出版以来，一直畅销50多年不衰。到2003年，该书已经销售一百多万册，至少翻译成了20种语言。《结构》确实改变了科学家从事科学探索的方式。虽然有些人不同意库恩的结论——事实上，其中一个长期的分歧被称为"科学战争"*——但《结构》吸引了科技界内外的读者，因为它捕捉到了20世纪60年代初期盛行于全社会的质疑精神。当时，"范式转移"似乎无处不在。所以，库恩关于转变科学观念的研究跟当时流行的讨论完全是一致的。

这个时期，社会学学科*（研究人类社会的历史和结构的学科）越来越盛行。社会科学家一直在考察社会条件如何影响观念，而现在他们又考察了社会条件是如何影响科学和科学史学家的。

这个时期，社会公众十分熟悉科学。很多人，诸如库恩的导师詹姆斯·布赖恩特·科南特，都认为科学是摆脱美国和苏联*之间冷战*紧张局面的关键。但是，越来越多的人在努力协调这个目标和核战争潜在毁灭之间的关系，人们质疑科学家们在开发这种毁灭性武器中所担当的角色。

《结构》对这些观念的发展起到了跳板的作用。人们从这本书中得出的有些结论让库恩本人甚感吃惊，有的甚至让他十分担忧。毕竟，他所写的是科学内在结构，而不是改变世界的思想力量。然而，毫无疑问的是，库恩的著作打开了质疑的大门，且这扇门不会再关闭。

《科学革命的结构》的学术价值

托马斯·库恩关于科学"革命"时期的思想，颠覆了数个世纪以来所形成的智慧。在库恩之前，科学家认为自己都是客观的观察者，把客观观察的结果和之前科学家的发现作为自己的研究基础。可自从库恩出版了《科学革命的结构》后，这些令人确信无疑的必然性不再适用了。

从这个意义来讲，《结构》本身就具有革命性。它改变了人们看待周围世界的方式——这个影响是如此之大，以致"范式"和"范式转移"成为日常用语。虽然库恩只是想把"范式"这个术语用于科学，但是学者们把它引入其他的学科，甚至商界也采用了这个术语。不出所料，"范式"现在出现在过度使用词语的清单上，那就变得毫无意义了。

虽然，库恩只是想用这本书来影响科学历史和科学哲学领域的读者，但是它也影响到其他领域的科学家、理科教师、对科学社会学感兴趣的学者，甚至普通大众。今天激进知识分子仍然熟悉库恩的思想。《结构》如今成为一些人的基础文献，他们是有兴趣对科学做后现代主义*批评的学者。粗略地说，这些方法通常采用的理论是，科学只是另一种受文化束缚的叙事方法。它们常常被用来展示历史上被边缘化的群体，如妇女、有色人种和殖民地被殖民者对科学发展产生影响的方式。同时，科学知识社会学*这门新学科认为，社会条件影响科学知识的创造——这是典型库恩式的见解。

虽然很多科学家一直不同意库恩的看法，但他们却承认他的这部著作对科学作出了巨大的贡献。其中一位是德裔美国科学哲学家卡尔·亨佩尔*，他虽然长期批评库恩，但同样承认这一点。亨佩尔在1993年写信给库恩，赞扬道："汤姆，不管你的同事采取什么立场，我相信他们都会对你富有争论性和启发性的思想感激不尽。"[2]

《结构》出版后半个多世纪来一直在启发并激励读者；即使在非科技界里工作的学者谈起"革命的结构"，也和库恩的定义一样。

---

1. N. M. 斯沃德罗："托马斯·S. 库恩传记回忆录"，2013年《美国国家科学院院刊》，登录日期2015年6月29日，http://www.nasonline.org/publications/biographical-memoirs/memoir-pdfs/kuhn-thomas.pdf。

2. 卡尔·亨佩尔："托马斯·库恩：同事和朋友"，载保罗·霍里奇等编《世界在变化：托马斯·库恩和科学的本质》，马萨诸塞州坎布里奇：麻省理工学院出版社，1993年，第7—8页。

第一部分：学术渊源

1 作者生平与历史背景

要 点

- 《科学革命的结构》对科学可以发现现实真相这一观念提出了持续的挑战。
- 托马斯·库恩做雷达技术专家的工作、读物理学专业研究生的学习经历,使他极不认可学者和公众对科学的基本理解。
- 库恩的工作受到第二次世界大战*和冷战*的影响(冷战指的是美国及其盟国和苏联及其盟国之间长期的、严峻紧张的政治局面)。

为何要读这部著作?

托马斯·库恩的《科学革命的结构》全面分析了科学知识是怎样发展的。库恩的这本书被誉为"20世纪最佳图书之一",过去半个世纪最畅销非小说类书籍之一。无论是在科学领域,还是在其他领域人类的探索中,此书改变了科学家们完成任务的方式,也改变了我们思考科学突进的方式。其核心概念——"范式"*,理解世界的智力模型;"科学革命"*,一种范式让位于另一种范式之时;"范式转移"*,一场科学革命之后的事态现状——已经成为流行词汇。

虽然人类认识的改变可以是革命性的这一观点已进入了日常语言[1],但是这部著作的核心问题——什么力量可以改变科学家思维和工作的方式?——今天仍然是激烈辩论的焦点。先前,科学革命的思想只是狭义地用在物理学中,库恩把这个术语广义地用在科学的各个领域；而他的读者则把它用来指代普通意义的智力探索[2]。

库恩的科学观是一种不断发展的知识体系。无论在他的祖国美国，还是在整个西方，它都影响了政治文化。他的结论证明了科学知识临时性和不确定性的本质，而他对科学进化的历史分析形成了科学史*这门学科；这表明科学知识随着时间的推移实际在发生变化，而不是固定不变、确信无疑的[3]。

然而，如果科学是临时性的，而不是确定性的，那么政治家和公众人物则再也不能把科学发现当作任何事物的绝对证据[4]。

> 一个更温和、更加支持科学追求的怀疑论并不排斥客观真实的可能性，而是对这种可能性持不可知论。托马斯·库恩就是这样的怀疑论者。
> ——迈克·W. 马丁：《创造力：科学的伦理和卓越》

作者生平

库恩大半生都做科学家。他上中学时就热爱数学和物理[5]，随后在哈佛大学*就读，到 1949 年他完成了物理学的本科和博士学业[6]。哈佛大学影响深远的校长詹姆斯·布赖恩特·科南特*，帮助库恩获得了历史和科学系的专业职位[7]。

第二次世界大战期间，科南特曾经是科学发展过程中一位重要的人物。作为"曼哈顿计划*"研发负责人，他不仅帮助研发原子弹，而且还说服美国总统哈利·S. 杜鲁门*，使用原子弹是不可避免的[8]。战争结束后，科南特重返哈佛大学校长岗位，着手一个项目，以便让科学拥有更多的听众——不仅在学术界，而且向社会公众公开。他招募了很多高级学者，如比利时流亡科学历史学家、化学家乔治·萨顿*，以及一些像库恩这样的初出茅庐者[9]。

库恩在其生涯里继承了科南特的事业，以公众和学者都能接受的方式投身于历史和科学哲学*的教学之中。1961年，他成为加利福尼亚大学科学史*教授，并著有《科学革命的结构》一书。1964年，库恩离任，旋即担任普林斯顿大学和同样享有极高声誉的麻省理工学院*的教授。他还担任了一年的科学史学会*主席[10]。

在20世纪60年代之前——也就是说，在《科学革命的结构》出版之前——科学家把科学史的写作当成主要工作之外的兼职[11]。库恩的作品改变了这一切，他让科学家有可能从物理学跨越到科学史和科学哲学。在这个过程中，他永远改变了全世界对科学的理解。

创作背景

在第二次世界大战期间，库恩从哈佛大学本科毕业后在部队服役两年，研究雷达技术。他的工作，使他对学术界以外的世界应用和实践科学的方式感到不满。战争结束后，他回到哈佛大学攻读研究生[12]。尽管他是学物理的正式学生，但他也想跨越科学在世界上的应用范围，去了解科学知识本身的基本原理[13]。

1945年，第二次世界大战末，美国政府通过展示巨大的军事力量——原子弹的爆炸，迫使日本投降；而库恩的导师科南特曾经帮助研发。原子弹摧毁了日本的两座城市：广岛*和长崎*。由此引发政治家和公众的疑虑，在战争中使用科学技术是否在道义上合情合理[14]。不管人们怎么看，事实已经引起公众的注意：科学发现有力量改变人类命运。

第二次世界大战结束后，冷战开始，一直持续到1991年前后。

冷战同样改变了库恩的科学观念。美国和苏联之间的竞争促进了科学技术的进步，也成为公开辩论的话题；库恩特别关注核武器的开发和可能爆发在这两个超级大国之间的核战争所带来的威胁。物理学在这些主要的研发中起着核心作用。加拿大科学史学者伊安·哈金*写道，在这个时代，"每个人都明白物理即行动所在[15]"——其中还包括库恩。物理科学及其基本原理和特征均已成熟，适宜分析，于是库恩的整个生涯便可以看成是对这个挑战的回应。

---

1. 托马斯·尼科尔斯："导言"，载托马斯·尼科尔斯等编《托马斯·库恩》，剑桥：剑桥大学出版社，2003年，第1页；伊恩·哈金："导读"，载托马斯·库恩著《科学革命的结构》第4版，伊利诺伊州芝加哥：芝加哥大学出版社，2012年，第XXXVII页。
2. 库恩：《结构》，第32页。
3. 克里斯多弗·格林："库恩走向何处？"，《美国心理学家》第59卷，2004年第4期，第271—272页。
4. 伊恩·哈金：《社会建设是什么？》，马萨诸塞州坎布里奇：哈佛大学出版社，1999年，第12页。
5. 亚历山大·伯德："托马斯·库恩"，《斯坦福哲学百科全书》（2014年冬季版），爱德华·N.扎尔塔等编，http://plato.stanford.edu/archives/win2012/entries/davidson/。
6. 伯德："托马斯·库恩"。
7. 斯蒂夫·福勒：《托马斯·库恩：我们时代的哲学历史》，伊利诺伊州芝加哥：芝加哥大学出版社，2000年，第9—11页。
8. 齐亚乌丁·萨达尔："托马斯·库恩和科学战争"，载理查德·阿皮尼亚内西编《后现代主义与大科学》，剑桥：艾肯图书公司，2002年，第200页。
9. 福勒：《托马斯·库恩》，第9—11页。

10. 伯德："托马斯·库恩"。
11. 杰夫·休斯："辉格党、普里格党和政治：当代科学编史问题"，载托马斯·苏德瑞编《当代科学技术编史》，阿姆斯特丹：哈伍德学术出版公司，1997年，第20—21页。
12. 伯德："托马斯·库恩"。
13. 玛尔妮·休斯-沃灵顿："托马斯·塞缪尔·库恩"，载《历史上五十位最重要的思想家》，伦敦：劳德里奇出版社，2003年，第188页。
14. 托马斯·C.里夫斯：《二十世纪美国简史》，牛津：牛津大学出版社，2000年，第137页。
15. 哈金："导读"，载库恩《结构》，第 IX 页。

2 学术背景

要点

- 科学史*和科学哲学*两个领域,解释了科学家是如何慢慢得出自己结论的。

- 20世纪60年代,学者们对如下观念提出了质疑:科学提供证据充分、准确无误的结果。科学实在论*与科学建构主义*对峙:科学实在论的支持者认为,科学家的意见不一定影响其结论;科学建构主义的支持者则认为,科学家并没有直接研究现实——他们只是研究通过实验收集的信息。

- 托马斯·库恩率先使用历史分析法来动摇逻辑经验主义*和逻辑实证主义*的基础——它们是以可验证性*、逻辑和理性主义*为基础的科学和哲学研究方法。

著作语境

托马斯·库恩的《科学革命的结构》属于科学史领域的著作:它研究科学知识如何随着时间的推移而发展,这一领域是由16世纪的几位一流科学家,如英国的弗朗西斯·培根*、德国的约翰尼斯·开普勒*等率先发起的。许多科学家用毕生的精力做出了意义重大的发现,这其中就有艾萨克·牛顿(一位物理学家,提出了对今天物理学研究仍然有影响的基本原则[1])。培根和开普勒意在用他们的科学史重建"可用历史",以此揭示他们毕生从事的"现代科学",是如何自早期"古代智慧"发展起来的[2]。

20世纪伊始,科学史同培根和开普勒时代相比并未发生多大的变化。例如，20世纪科学史家把科学知识看成是事实和数据的积累。随着人们把科学实验加入事实的储存之中，他们发现科学具有线性轨迹特征。所有新的科学发现都建立在先前发现的基础上，以便取得科学知识的"进步"。此外，科学家只是被动地观察并记录其结果，而不在其发现之中添加个人的意见和偏见[3]。

> 库恩通过对历史的考察，探讨了理性和证据的哲学问题……一方面，逻辑经验论者明确区分了科学历史和科学心理的问题，另一方面又区分了证据和理由的问题。而库恩则有意把逻辑经验论者坚持分开的东西融合在一起。
> ——彼得·戈弗里-斯密斯：《理论与现实：科学哲学导论》

学科概览

库恩于1962年撰写《结构》之时，科学史见证了重大的变革性发展。先前科学史是由科学家撰写的，而现在是由历史学家撰写；甚至连那些未接受正规科学训练的历史学家，也对科学史感兴趣[4]。比利时裔美国科学家和历史学家乔治·萨顿*力促这种变革，他说："科学史学家应该了解历史**和**科学……光有善意远远不够[5]。"

研究科学的历史学家借鉴了新兴的历史分支学科：思想史*。在美国，思想史这个领域是由德裔美国哲学家和历史学家亚瑟·O.拉夫乔伊率先开辟的，揭示了几个世纪以来思想是如何演变以表达不同内容的。拉夫乔伊"孜孜不倦地开拓这门学科，将其构建成纯粹的认知世界观*及其组成'单元思想*'的研究"[6]。

换言之，拉夫乔伊认为，包括科学家在内的人们通过形成心理意象，即"认知世界观"，来回应世界。他们利用这些心理意象把大脑中孤立的、他们称之为"单元思想"的概念集合联系起来。有关世界的任何观念，例如，不管是科学的、引力的还是死亡的，都会随着时间的推移而发生变化，因为人们往往会用不同的方式把单元思想联系在一起。所以不同的人、地方、时代，都会对科学、引力和死亡——或者他们构建于单元思想的别的任何东西，有不同的观念。

学术渊源

库恩所受的影响来自科技、社会科学和科学史等领域里的学者，他也得益于这些学科之间的交流。

当科学家、哈佛大学校长詹姆斯·布赖恩特·科南特*开始对公众进行科学教育时，他寻求了库恩的帮助。库恩在研究公共科学教育时，发现了开普勒和培根曾经描述的"现代科学"和"古代智慧"之间的差异[7]。科南特还把库恩介绍给萨顿——科学教育项目的另一个成员——库恩学习了萨顿对科学史的创新思维方法[8]。

社会科学、心理学和语言学界的学者，激励库恩以新的方式思考人的思维和语言是怎样塑造世界观的。

法国心理学家让·皮亚杰*关于儿童发展的理论和格式塔心理学流派*的实验均强调，人们用理论框架来分析客观世界。库恩接受了这一观点[9]。换言之，他承认思维过程离不开我们对世界的认识。从语言学角度看，库恩学会了语境改变词义的方式，这加深了他对皮亚杰心理学的理解：每个个体都有不同的视角。

然而，对库恩影响最大的是奥裔英国哲学家路德维希·维特根斯坦*。维特根斯坦认为，个体表述对世界的认识实际是"诠释"，"当我们诠释时，我们形成了假设，而这些假设可能是错误的[10]。"换言之，即使人们在任何特定时刻所说的是真的，人们对世界的认识也会不正确。

维特根斯坦深受18世纪德国哲学家伊曼努尔·康德*的启发，创立了自己的语言理论。库恩也认真研究了康德。康德发现了人们分析事物的四个范畴：数量、质量、关系和形态。康德通过这些范畴，依据客观对象自身的特征，并参照其存在的环境，来描述客观对象。以往的哲学家只关注其中一两个；[11]也就是说，康德的四个范畴使他形成了一种方法，来解释物理对象是如何表现的。

库恩也想依据基本概念解释科学活动。康德的范畴影响着维特根斯坦20世纪的理论，即我们是如何利用语言来解释世界的。引申开来，康德的思想向库恩揭示了日常语言如何影响我们对世界的理解。

科学史学家启发了库恩对文化假设在科学结论中的作用的关注。移民到美国的法裔俄国学者亚历山大·柯瓦雷*重新提出了以下观点，即科学知识是由于科学家的思维方式发生了变化或革命而发展的[12]。库恩还发现，法国科学哲学家海琳·梅斯热*关于17世纪化学的论著和德国历史学家安内利泽·迈尔*中世纪科学的论著引人深思。但是也许对库恩来说，最重要的灵感之源是隐性知识*（译者注：也有人把 tacit knowledge 翻译成"意会知识"）（大致指的是这样一种知识，不能轻易通过简单的写或说在人和人之间转移），这是匈牙利裔英国学者迈克尔·波兰尼*提出来的，他在很多学科领域里都是出类拔萃的。波兰尼的思想鼓励了库恩去研究文化态度以及科学方法，是如何影响科学家的结论并在他们的结论中被觉察到的[13]。

1. A.鲁珀特·霍尔:《1500—1750期间的科学革命》,伦敦:朗文出版公司,1983年,第134、143页。
2. 霍尔:《1500—1750期间的科学革命》,第18页。
3. 托马斯·S.库恩:《科学革命的结构》第4版,伊利诺伊州芝加哥:芝加哥大学出版社,2012年,第31页。
4. 杰夫·休斯:"辉格党、普里格党和政治:当代科学编史问题",载托马斯·苏德瑞编《当代科学技术编史》,阿姆斯特丹:哈伍德学术出版公司,1997年,第20页。
5. 乔治·萨顿:《科学史指南》,马萨诸塞州沃尔瑟姆:克罗尼卡波特尼卡出版公司,1952年,第IX—X页。
6. 安德鲁·朱厄特:《科学、民主和美国大学:从内战到冷战》,剑桥:剑桥大学出版社,2012年,第257页。
7. 托马斯·库恩:《结构以来的路:1970—1993哲学论文集与自传式访谈》,詹姆斯·科南特和约翰·海于格兰编,伊利诺伊州芝加哥:芝加哥大学出版社,2000年,第16页。
8. 史蒂夫·福勒:《托马斯·库恩:我们时代的哲学历史》,伊利诺伊州芝加哥:芝加哥大学出版社,2000年,第9—11页。
9. 库恩:《科学革命的结构》"序言",第XI页。
10. 路德维希·维特根斯坦:《哲学研究》第3版,牛津:布莱克维尔出版公司,2003年,第181页。
11. 米凯拉·马斯米:"康德之后的哲学与科学",载安东尼·奥赫尔编《哲学概念》,剑桥:剑桥大学出版社,2000年,第282页。
12. 保罗·霍伊宁根–休内:《重构科学革命:托马斯·S.库恩的科学哲学》,伊利诺伊州芝加哥:芝加哥大学出版社,1993年,第XIX页。
13. 亚历山大·伯德:《托马斯·库恩》,切舍姆:欧酷曼出版社,2000年,第14—20页。

3 主导命题

要点

- 科学史学家质疑,科学家是否真正客观地工作,是否不受个人观点和职业立场影响去研究现实。
- 逻辑经验论者*认为,科学家通过科学数据证实自己的所有发现;但是科学史家则认为,证据这个概念本身是随着时间的推移而发生变化的,因而实证是相对的。
- 库恩把这些观点结合在一起,利用实证和相对性形成对科学知识的新理解。

核心问题

在《科学革命的结构》中,托马斯·库恩试图"阐明并加深对**当代**科学方法或展示这些方法发展概念的理解"[1]。换言之,他重塑了16世纪形成科学史这门学科以来,科学史家提出的中心问题:科学家是如何认识真实世界的?

这一中心问题可以分解为两个部分:一部分是关于**科学家**的问题,另一部分是关于**现实自身本质**的问题。史学家质疑科学家在评估数据时是否能摆脱个人和文化的偏见,现实是否独立存在于负责评估的科学家的思维之外。

这些疑问揭示了一个基本问题,该问题使历史学家和科学家之间持续存在差异[2]。一方面,"现实主义者"*认为,科学家有机会接触由客观对象和物质组成的真实世界,对它进行分析、实验和描述。通过这个方式,现实主义者相信科学家能够揭示真理,而不把个人意见掺杂其中妨碍研究³。世界以其本来面貌客观存在,显而易见,所有科学家都可以亲眼目睹。另外的学者——"建构主义者"*——则持相反的观点。他们认为,科学家构建了自己的现实图景,并不由自主地在其中融入了自己的思想、认识和直觉。他们自己的研究,他们在科学界的地位,连同大量智力和社会可变因素,也影响他们对现实的认识⁴。

> 现在有一种日益增长的趋势,即把知识理解为一种历史经验的构建,并挑战自然科学与人文科学及现实生活中客观性之间的任何原则性分裂。赞成这些主题,实际上就是否认任何偏向自然科学的特权或三六九等的知识。
> ——约瑟夫·马格利斯:《作为问题的客观性》

参与者

在库恩出版《科学革命的结构》之前,现实主义和建构主义的立场都有自己的代表:科学史家采取建构主义立场,而逻辑实证论者或经验论者*(其研究方法"可以概括为知识仅来源于经验")⁵则采取现实主义立场。

1960年以前,逻辑经验主义*——科学哲学的一个流派*,起源于17世纪英国哲学家约翰·洛克*发轫的传统,由诸如18世纪苏格兰哲学家大卫·休谟*、19世纪德国哲学家恩斯特·马赫*传承下来——似乎更有说服力。在第一次世界大战前后,德国哲学家鲁道夫·卡尔纳普*、奥图·纽拉特*和卡尔·亨普尔*觉察到哲学规范下降,他们的工作——后来被称为逻辑经验主义——是对此的回应⁶。逻辑经验主义通过增加两个中心思想,即分析与综合的区分思想*和意义的可验证性*,为洛克建立的古典经验主义传统增添了新的内容。

德国哲学家伊曼努尔·康德*在17世纪创立了分析与综合区分的思想。一个分析命题*，一定是诸如"所有单身汉均未婚"这样的陈述。单身汉的定义——未婚男子——本身就包含在命题之中。相比之下，一个综合命题*如"所有单身汉均不幸福"，并不包含其定义。"不幸福"并不是"单身汉"定义的一部分，而且这个陈述对有些单身汉来说是真实的，对另外的单身汉却未必如此。逻辑经验主义者把数个世纪的古老概念，同实证主义原则*新概念匹配在一起——实证主义认为，只有两种陈述才具有实际意义：一种是具有逻辑必要性的陈述，另一种是可以通过实验得到证明的陈述。根据实证思想，第二位科学家只有通过重复其实验，来检验第一位科学家对整体的认识，以确保其分析与综合的有效性[7]。

逻辑经验主义者把分析与综合区分的思想同实证主义理论结合在一起，认为科学家分析并思考与**部分整体**有关的数据，利用理性这一过程感知经验世界，从而逐渐理解**整体**[8]。

逻辑经验主义者认为，这些基本概念任何时候都可以运用在所有科学里。其结果是，科学不断积累新知识和新发现，就像它对16世纪科学史的创造者*（英国哲学家弗朗西斯·培根*和德国天文学家约翰尼斯·开普勒*）而言那样。

相比之下，历史学家采取建构主义立场。正如他们所看到的，科学家是按照自己时代特有的原则行事的。纵观几个世纪，他们发现科学以一种不规则、不可预测的方式发展，并坚持认为，要重建精确的科学史，人们需要把科学活动"置于社会背景中考虑"。

在20世纪中期，科学史学家指出了19世纪晚期法国哲学家和历史学家皮埃尔·迪昂*作品的重要性。迪昂认为，学者们需要"理性重建（科学史，其中）只包含与哲学相关的内容"[9]。换言之，历史学家应该研究科学家的信仰以及科学家怎样按照信仰行事。科学史家如果想了解科学是怎样随着时间的推移而渐渐发展的，就应该研究这一点，而不应该像逻辑经验主义者那样运用普通哲学概念。

当代论战

20世纪60年代，当库恩撰写《科学革命的结构》的时候，逻辑经验主义者和历史学家之间的辩论愈发胶着，双方出现了交叉重叠，库恩从双方都汲取了一些元素。

双方都基于对语言在创造知识中所起作用的具体理解。逻辑经验主义从语言学家的著作中，得出实证理论和其对分析与综合的区分。除了康德以外，他们还借鉴了20世纪哲学家路德维希·维特根斯坦*的哲学思想。虽然他们两人相隔几个世纪，但都强调了潜在的心理操作是理解个人获得知识的关键[10]。库恩对康德的兴趣，加强了他与逻辑实证主义者的联系；他们都想把自己对科学的普遍理解，建立在具有逻辑基础的基本原理上[11]。

虽然库恩很少公开把自己同逻辑经验主义者联系在一起，但他确实把自己与建构主义者对科学的语境解释联系在一起。实际上，他认为，他把《科学革命的结构》当成反对20世纪50年代主导科学教材的逻辑经验主义的声明[12]。他声称，他通过提出科学是在无序的循环中运作这一观点（所以科学知识发展是不均衡的，这一观点与逻辑经验主义的假设形成了鲜明的对比）来表明他的立场。他把自己的观点叫做"科学研究中的史学编撰革命"，这一术语引自迪昂、亚历山大·柯瓦雷*以及其他许多学者，以便促进对科学的历史理解而不是哲学理解[13]。

科学史兴起于迪昂和法裔俄籍科学史家亚历山大·柯瓦雷的研究工作。柯瓦雷创造了"科学革命"这个词语,库恩在其《结构》里始终对他盛赞有加。正是这种科学史把库恩领上历史诠释的道路。在库恩看来,科学史在历史学和科学两大部门里均弱化为边缘学科,但是柯瓦雷的研究工作夯实了它在学术上的可靠性[14]。库恩认为,柯瓦雷另辟蹊径,用新的方法撰写历史课本,以便能"揭示科学新形象的可能性"[15]。换言之,库恩相信,历史学研究的发现能够为理解科学运作带来一场革命。一场当代激烈的辩论正在猛烈地进行,随着《结构》的出版,库恩在这场辩论中占据了中心位置。

---

1. 托马斯·S.库恩:"科学史",载《必要的张力:科学传统和变革研究文选》,伊利诺伊州芝加哥:芝加哥大学出版社,1977年,第107页。
2. 彼得·格弗里-斯密斯:《科学哲学导论:理论和现实》,伊利诺伊州芝加哥:芝加哥大学出版社,2003年,第5页。
3. 伊恩·哈金:《什么社会建设?》,马萨诸塞州坎布里奇:哈佛大学出版社,1999年,第68页。
4. 格弗里-斯密斯:《理论和现实》,第6页。
5. 格弗里-斯密斯:《理论和现实》,第228页。
6. 格弗里-斯密斯:《理论和现实》,第28页。
7. 格弗里-斯密斯:《理论和现实》,第28页。
8. 格弗里-斯密斯:《理论和现实》,第27页。
9. R. N. D. 马丁:《皮埃尔·迪昂:有信仰的物理学家工作中的哲学与历史》,伊利诺伊州拉萨尔：公开法庭出版公司，1991年，第139页。

10. 托马斯·S.库恩："后记"，载保罗·霍里奇编《世界在变化：托马斯·库恩和科学的本质》，马萨诸塞州坎布里奇：麻省理工学院出版社，1993年。
11. 托马斯·S.库恩：《结构以来的路：1970—1993哲学论文集与自传式访谈》，詹姆斯·科南特和约翰·海于格兰编，伊利诺伊州芝加哥：芝加哥大学出版社，2000年，第264页。
12. 托马斯·S.库恩：《科学革命的结构》第4版，伊利诺伊州芝加哥：芝加哥大学出版社，2012年，第2页。
13. 库恩：《结构》，第3页。
14. 库恩：《结构》，第3页。在1971年的一篇文章中，他谈到柯瓦雷是他的"领班"，地位高于其他所有历史学家；托马斯·S.库恩："物理学发展中的原因概念"，载《必要的张力》，第21页（原文是用法语撰写的，这是出版的翻译版本）。
15. 参见库恩：《结构》，第3页。

4 作者贡献

要点

- 托马斯·库恩认为科学沿着以下路线发展：预共识科学*→常规科学*→危机科学*→非常规科学*→科学革命*。
- 《科学革命的结构》通过提出这种周期性的替代方案，颠覆了科学领域线性和累积进步的主流观点。
- 尽管周期模型发展了科学革命的旧概念，但其独创性在于它对客观性提出的挑战（即，假定科学家能够在没有自己世界观干预的情况下考察"现实"）。

作者目标

在《科学革命的结构》里，托马斯·库恩的目的是展示科学知识是循环发展的——这是他攻读博士学位期间就开始形成的理论，当时他写道："接触了过时的科学理论和实践，从根本上破坏了我对科学本质的基本理解。"[1]

这种周期性的理解，始于承认科学史学家用与时俱进的概念来构建他们的历史[2]。如果历史的分析可以是周期性的，那么知识本身亦可以。

库恩阐述的周期从"预共识科学"——对立的思想流派从不同的角度讨论同一个对象时，得出不同的结论[3]——开始，经过从不同的阶段，直到科学革命时期，再重返回去。对库恩来说，预共识科学不太算科学[4]，因为没有任何共同的观点存在，任何想法都可以尝试。他举了一个例子，在英国物理学家艾萨克·牛顿*的研究之前，17 世纪科学家对光波传播有不同的解释——这是物理学中一个叫光学的领域[5]。

"常规科学"从"预共识科学"中产生，其中一个思想流派赢得所有科学家的拥护；这些科学家于是在科学实践的公认规则范围内，以协商的方式修改既定概念[6]，并以线性的方式积累知识[7]。

常规科学让位于危机科学和非常规科学，并最终让位于科学革命。一旦科学家面临常规科学时期出现的费解现象，再也无法解释世界时，科学的危机就出现了。库恩称之的"危机状态下的科学[8]"，重新审视指导常规科学的惯例和既定概念。当危机发生时，几个范式*为被科学共同体*接受而竞争。依据库恩的观点，这些范式是不可通约*的——即它们是不同的世界观，对现实的解释差异如此之大，以致不能共存。

库恩认为，正是科学家的性格和爱好，最终解释了科学共同体决定接受何种范式。如果说范式对世界做出不同解释，而依据各自论点，每一种解释都言之有理，那么对各种范式选择的理由并不是科学的，而是个人的[9]。

这些危机扰乱了常规科学的线性发展过程，使科学发展不平衡。它们并不总是以革命告终，而通常会找到一个解决方案。也许一种解决方案来自"常规科学"阵营，也许科学界愿意搁置问题，留给后代去解决。然而，一旦这些解决方案都失败，就会发生革命。革命结束后，科学便又回到"常规科学"状态，整个过程又开始循环。

如果某一科学领域达到革命阶段，其影响是巨大的。革命产生新的范式，而范式需要彻底"重建"这一科学领域[10]。到革命结束时，这个行业（译者注：指科学研究）本身将会改变其观点、方法和目标[11]。

在提及心理学理论"格式塔*"时——这个理论假设,我们在大脑里把我们的经验进行排序来构建一种世界观。以这样的方式,使某一单独的、新的经验能够彻底改变我们对何为"真实"的理解——库恩把这种变化比作一种视觉格式塔,凭此曾经视为鸟类动物现在却视为羚羊类动物[12]。革命范式的转变*,在科学家的概念体系中产生了同等的位移[13]。

> 《科学革命的结构》的起源解释了,为什么这里所提出的问题与最近皈依科学史的人提出的问题是相仿的。这些人接受过科学家的培训,也接受过受到社会学强烈影响的历史学家的培训;事实上,库恩认为这本书是对知识社会学的贡献。
>
> ——玛丽·博厄斯·霍尔:评《科学革命的结构》

研究方法

库恩试图从科学背景中看待科学,就像科学家自身解释的那样。他说,他试图通过考虑一组科学家们"在某个特定时期从事某一特定科学专业时"的所思所想,来"展示这种科学在当时的历史整体性"[14]。

虽然库恩高度重视历史,把它当作分析科学活动的出发点,但这并不意味着其研究方法所要达到的目的。从对科学家如何工作进行历史学理解中,他坚持认为他能够推断出一系列的概念,来解释科学家怎样工作、为什么工作,以及他们如何继续工作[15]。

这种方法与逻辑经验主义的方法*形成鲜明的对比。后者认为无法证实的科学实验对现实的证明作用是微乎其微的。因此,库恩的思想和逻辑经验主义思想之间的关系仍然不清晰，库恩本人经常回避这种比较[16]。

库恩在很大程度上吸收了20世纪初期历史学家的研究成果，如法裔俄籍的亚历山大·柯瓦雷*（"科学革命"这个术语的创造者）和法国历史学家皮埃尔·迪昂*。这些学者试图通过仔细研究科学史上的诸多事件来了解科学家实际上是如何工作的，并由此揭示科学的真正特征。

逻辑经验主义首先借助哲学概念来解释历史科学，而库恩则有效地反其道而行之，首先用历史揭示哲学概念："库恩的研究工作似乎表明，将科学的哲学问题与科学史问题联系起来是多么有趣。[17]"

时代贡献

《科学革命的结构》的独创性在于以下事实：这本书建立在对科学史现有的理解框架之上——特别是认为科学在人类智识产业中占有特殊地位这一观点——并同时动摇了这些相同的框架，认为科学进步是周期性的，科学家在界定"现实"时有很大的发言权。当那些杰出的逻辑经验主义者试图收集关于科学的性质、功能和目的[18]最完整可靠的阐述时，库恩主动让他们在其项目——《统一科学百科全书》之中收录自己的著作《科学革命的结构》。

库恩著作另外一个极具独创性的方面是，它颠覆了逻辑经验主义者和历史学家使用的现有概念。库恩展示了在某些显而易见、持续不变的观点中包含着彻底重释这些观点的种子——这与逻辑经验主义者收集关于科学的完整可靠的阐述的理想截然相反。库恩的这一观点震惊了科学界。

1985年，美国哲学家阿瑟·丹托*说，库恩确保了"后经验主义"的来临。他写道:"确实存在科学统一体,从某种意义上说,所有科学都是在历史的框架下形成的,而不是像以前那样相反,历史被置于科学框架之下,用物理模型来解释[19]。"

因此,库恩的独创性在于通过揭示科学固有的不统一来创造统一性,从而使科学超越逻辑经验主义。

---

1. 托马斯·S.库恩:《科学革命的结构》第4版,伊利诺伊州芝加哥:芝加哥大学出版社,2012年,第V页。
2. 库恩:《结构》,第2、7页;库恩:《必要的张力:科学传统和变革研究文选》,伊利诺伊州芝加哥:芝加哥大学出版社,1977年,第XI页。
3. 库恩:《结构》,第4、12—13、47—48、61—62、178—179页。
4. 库恩:《结构》,第13—18页。
5. 库恩:《结构》,第16页。
6. 库恩:《结构》,第17—19、178页。
7. 库恩:《结构》,第52—53页。
8. 库恩:《结构》,第82—87、101、154页。
9. 库恩:《结构》,第95页。
10. 库恩:《结构》,第85页。
11. 库恩:《结构》,第85页。
12. 库恩:《结构》,第85页。
13. 库恩:《结构》,第102页。
14. 库恩:《结构》,第3页。
15. 库恩:《结构》,第3页。
16. 约瑟夫·劳斯:"库恩的科学实践哲学",载托马斯·尼科尔斯编《托马斯·库恩》,剑桥:剑桥大学出版社,2003年,第101页。
17. 彼得·格弗里-斯密斯:《科学哲学导论:理论和现实》,伊利诺伊州芝加哥:

芝加哥大学出版社，2003年，第78页。
18. 彼得·诺威克：《崇高梦想："客观问题"和美国历史行业》，剑桥：剑桥大学出版社，1988年，第526页。
19. 阿瑟·丹托：《叙事与知识》，纽约：哥伦比亚大学出版社，1985年，第XI—XII页。

第二部分：学术思想

5 思想主脉

要点

- 《科学革命的结构》的主题是：科学革命*、范式*和不可通约性*。
- 库恩认为，当科学家挑战并颠覆范式时，他们推动科学进步，从常规科学*（即按照普遍的世界观进行科学实践）到危机和科学革命，再到新的常规科学。
- 这一论点，作为简要分析首次出现在《国际科学百科全书》的两卷中。

核心主题

在《科学革命的结构》中，托马斯·库恩根据以下三个主题构建了自己对科学的独特见解：

- 科学革命的思想
- 范式的概念
- 不可通约性的概念

正如库恩所见，"科学革命"发生在重大突破推动科学进步之时，这些重大突破永久性地改变科学家理解世界的方式。

库恩使用"范式"这个词来指代科学家从事研究时所持的核心概念，以此加深他们对世界的认识。库恩对科学的理解牢牢扎根于这个核心思想。他把范式定义为科学行为"公认的模型或模式"[1]。

"不可通约性"这个概念与范式紧密相连，形成了库恩核心思想的最后一部分。不可通约性描述了非常规科学*时期范式之间的关系，这个时期异常*现象（即不"适合"公认模型的观察）使人们对共识范式*产生怀疑。当两种或者更多的范式竞相对现实作出不同的解释而出现危机时，范式本身变得不可通约——即持不同范式的科学家无法相互交流。即便可以交流，他们也无法相互理解各自的发现，因为他们根据各自喜欢的范式采用不同标准的证据或者不同的概念。其结果，这种不可通约性和混沌持续下去，直到达成一种新的共识范式[2]。

这三种思想结合在一起，使库恩能够为科学的进步提出自己的观点。他认为这是周期性的，也是由科学家群体*决定的。因为科学家通过他们的行动，既接受了范式，又推翻了范式。

库恩把自己的观点建立在科学重大突破的历史范例上，从而使其思想合理化。在考虑过去的科学进步时，他说明了自己的通用概念如何解释和帮助我们理解具体的发现。其目的，是让自己的工作最终质疑对科学的线性与累计性的理解，这种理解是科学哲学*的逻辑经验主义*的方法特征。

> 科学被认为是一套解决特定领域技术难题的工具，随着时间的流逝，其精度和广度明显提高。作为一种工具，科学无疑是进步的。
>
> ——托马斯·库恩：《科学革命的结构》

思想探究

库恩对科学进步历史解读的关键，在于科学革命这一概念。其观点是"每一次科学革命都会改变经历这场革命的（科学家）群体的历史观"，其结果是"改变了后革命时期教科书和科研出版物的结构"[3]。换言之，每当科学出现重大突破时，科学家便用新发现的知识重写科学历史。在库恩看来，科学知识并非积累的，重大突破也不能增加现有知识，只能产生"世界变化"，以便科学家能在其中用不同的方式看待现实[4]。

在任何科学革命中，科学家必须建立适当的范式，以便解释他们为什么从事科学研究。出现一种范式，是为了解释并指导科学家来研究"一些问题，而从业者开始认识到这些问题是十分严重的"[5]。根据库恩的首席翻译和支持者，即德国哲学家保罗·霍伊宁根-休内*的观点，范式通过二分过程获得认可。首先，科学家认为一种情况引发一个科学问题。然后，他们就一个范式达成共识，这个范式在科学上能提供一个公认的问题解决方案[6]。范式既能解释科学问题，又能指导科学实践[7]。虽然库恩对范式的描写有些"刻板"，认为是常规科学的一部分，而常规科学本身具有"刻板"特征[8]，但是它们并不是死板得让科学家终止"解谜"研究[9]。

科学家会继续研究工作，并遇到新的问题。事实上，范式必须是相对开放的，因为它们可以提供未来研究的方向，帮助科学家决定哪些事实值得搜集和分析[10]。一旦科学家遇到了范式不能解释的问题，不可通约性（即不可能使两个相反的范式协调一致）这个概念则解释了在危机科学时期，甚至是科学革命本身，如何考虑和抛弃潜在的替代共识范式。

正如我们所看到的，对库恩来说，范式根本上是"公认的模型或模式"[11]。因此，社会对科学家群体的认可，对于建立一种范式是至关重要的。库恩看清了范式和规则之间的根本区别，对他来说，科学界*的规则是次要的、简化的——因此并不是和范式一样有效。规则对于科学事业来说过于僵化。在范式概念化过程中，库恩呼吁"会意知识*"，这一概念是由英裔匈牙利籍科学哲学家迈克尔·波兰尼*提出来的。会意知识是"通过实践获得的知识……是不能明确表达的"——换言之，是科学家公认的不言而喻的准则[12]。

库恩相信科学范式有自己的生命周期，它们始于解决一种特别严重的问题，终于未能解释随后出现的同样重要的异常现象。然而，范式不会就此简单地消失，它们抵制变革，直到一种新的范式取而代之。一种范式取代另一种范式的过程，出现在他称之为科学革命的时期——整个科学界的全面更新。

这导致了库恩的相关论点，即不可通约性。不可通约性描述了范式之间的关系。在库恩看来，只有一种范式才能够在特定的时间里存在于具体的科学研究领域，范式拥有绝对的权威，决定了科学学科的全貌。当一种范式取而代之，"这个行业将会改变本领域的观点、方法和目标[13]。"一旦科学家采用一种新的范式，世界的整个"概念系统"在这个学科中发生了变化；科学家在一个"新的世界"里工作[14]。库恩把范式之间的竞争比作不同革命派系之间的战斗，获胜方基本是大获全胜。

为了解释库恩的不可通约观点，设想"A"代表"世界"，"B"代表人类观察者。

库恩声称，科学家在开始观察过程之前就采用了各自的范式。所以，当观察者 B 看到世界 A 时，她或他通过自己已经采用的范式认识到了世界形态。如果 B 采用范式 1，观察者的行动则产生世界 A1；但是如果 B 采用范式 2，观察者则产生世界 A2。因为，解释甚至在观察行为之前就开始了。然而，范式并不决定任何一个世界的存在，世界的现实就是观察者的独立性。

许多读者理解到，不可通约性是一种全新的、完全的、瞬间的"世界变化"[15]。这种理解引起了危机,如澳大利亚哲学家霍华德·桑克指责库恩怀疑"某种形式的唯心主义*"[16]。唯心主义认为,人类思维在影响人们如何看待世界中起着决定性的作用——人们并不是直接接触现实的。但这没有描绘出库恩的立场:"虽然这个世界并没有因为范式的改变而改变",他写道,"范式转换后科学家却在一个不同的世界里工作。"[17]

在这样的陈述里,库恩解释说,他的意思并不是指现实依赖观察者,而指的是现实的感官知觉依赖之前建立的范式。在他看来,科学恰恰有能力把理论和知觉结合到客观、"真实"的自然界之中[18]。

然而,库恩似乎故意表现得模棱两可。他意识到存在一个"真实"世界;它离人类知识如此遥远,以至于几乎无关紧要[19]。他列举了以下例子,18世纪法国化学家、氧的发现者安托万·拉瓦锡*的研究工作引起了一场化学革命,形成了全新的观察,即"既然我们没有理由假定自然是固定不变的,变的只是拉瓦锡的看法,按照思维经济原则,我们就应该说:发现氧气之后,拉瓦锡是在一个不同的世界里工作。"[20]

然而,对桑克来说,这些陈述表明在库恩眼里,"真实"世界是"可有可无的",或者"无关紧要的",因为它通过科学感知或信仰根本不可触及[21]。库恩的立场(桑克认为)基于这种思想,即科学家不可能接触到真实世界,科学家及其范式所属的科学共同体影响了他们的思维。

德国哲学家保罗·霍伊宁根–休内提出了一个更加微妙的解释,区分了世界本身("真实"世界)和现象世界(人类经验世界)[22]。真实世界不依赖科学,而现象世界由于受到范式的影响,则依赖科学——这个观点起源于德国哲学家伊曼努尔·康德*，库恩承认其思想受到康德的影响 23。

虽然库恩赞同霍伊宁根-休内的康德式诠释，但是英国哲学家亚历山大·伯德*仍然持怀疑态度。对库恩来说，把他的工作跟康德的联系在一起，试图让"他早期的思想拥有（一种特定的）哲学上的成熟，但这种成熟实际上并不存在。"24

语言表述

库恩在《科学革命的结构》一书中，通篇以不同的方式使用"范式"和"不可通约性"这样的词语，认为"范式概念将经常取代已经熟悉的种种概念" 25。

科学哲学家玛格丽特·玛斯特曼*在《科学革命的结构》中确认了"范式"这个术语有 21 种意义 26——一种语言上的模糊性，既有优点，也有缺点。一方面，库恩的范式思想本身模糊，很难为科学哲学家所接受；另一方面，库恩能把许多相互联系的概念凝练成一个词，这将对学术界和大众文化产生重大影响。

同样，霍伊宁根-休内注意到，库恩用两种不同的方式使用"不可通约性"这个术语。起初，随着科学从前革命时期发展到后革命时期，它出现在库恩讨论问题和标准的变化方式中 27。但是，当库恩在其后期工作中重新启用这个概念时，他以一种更加激进的方式使用这个术语。他说："科学家对环境的感知必须重新训练……他所探究的世界似乎各处都会与他以前所居住的世界彼此间不可通约了。"28

库恩把这种现象同"格式塔转换*"作比较，"格式塔转换"是心理学的格式塔学派的一个术语。根据这个术语，一些新的、个别的感知信息可以导致我们对所感知事物的理解发生根本性转变。他还援引了 16 世纪尼古拉·哥白尼公布其宇宙模型后科学界发生的革命性变化;"哥白尼之后",他写道,"天文学家生活在一个不同的世界里。"[29]

---

1. 托马斯·S.库恩:《科学革命的结构》第 4 版,伊利诺伊州芝加哥:芝加哥大学出版社,2012 年,第 23 页。
2. 库恩:《结构》,第 85 页。
3. 库恩:《结构》,第 VIII 页。
4. 库恩:《结构》,第 19 页。
5. 库恩:《结构》,第 24 页。
6. 保罗·霍伊宁根-休内:《重构科学革命:托马斯·S.库恩的科学哲学》,伊利诺伊州芝加哥:芝加哥大学出版社,1993 年,第 134—135 页。
7. 库恩:《结构》,第 46 页。
8. 库恩:《结构》,第 19、49、64 页。
9. 库恩:《科学革命的结构》,第 49 页。
10. 库恩:《结构》,第 25—26、48 页。
11. 库恩:《结构》,第 23 页。
12. 库恩:《结构》,第 44—45 页。
13. 库恩:《结构》,第 85 页。
14. 库恩:《结构》,第 102 页。
15. 汉纳·安德森、彼得·巴克、陈向:《科学革命的认知结构》,剑桥:剑桥大学出版社,2006 年,第 106 页。
16. 霍华德·桑克:"库恩不断变化的不可通约性概念",《英国科学哲学杂志》第 44 卷,1993 年第 4 期。
17. 库恩:《科学革命的结构》,第 121 页。

18. 库恩:《结构》,第 134 页:"it is hard to make nature fit a paradigm.(要使自然界符合范式是很困难的。)"
19. 库恩:《结构》,第 111 页。
20. 库恩:《结构》,第 118 页。
21. 桑克:"库恩不断变化的不可通约性概念",第 764 页。
22. 霍伊宁根-休内:《重构科学革命》,第 239 页。
23. 托马斯·S.库恩:《结构以来的路:1970—1993 哲学论文集与自传式访谈》,詹姆斯·科南特、约翰·海于格兰编,芝加哥:芝加哥大学出版社,2000 年,第 264 页。
24. 亚历山大·伯德:"科学革命的结构及其意义:五十周年纪念版论文评述",《英国科学哲学杂志》第 63 卷,2012 年第 4 期,第 869 页。
25. 库恩:《结构》,第 11 页。
26. 玛格丽特·玛斯特曼:"范式的本质",载伊姆雷·拉卡托斯和 A. 马斯格雷夫编《知识的批评与增长》,剑桥:剑桥大学出版社,1970 年,第 59—89 页。
27. 库恩:《结构》,第 103 页。
28. 库恩:《结构》,第 112 页。
29. 库恩:《结构》,第 117 页。

6 思想支脉

要点

- 《科学革命的结构》对科学共同体*的社会学*(即从社会角度研究科学世界)、相对主义*的概念(大致意思是,相信在科学中是不可能有"最终"答案的)和自然科学的多样性做出了宝贵的贡献。
- 这些思想源自库恩关于科学进步的周期性结构的总体论点。
- 虽然有时相对主义被忽视了,但是总的来说,它对社会科学家产生了最广泛的影响。这项工作也改变了科学史学家思考科学共同体的社会学和自然科学多样性的方式。

其他思想

托马斯·库恩《科学革命的结构》中的主要思想包括科学革命*、范式*和不可通约性*——这些思想扣人心弦,促使科学史*突飞猛进地发展。此外,这本书还包括三个次要思想,每一个都强化了库恩关于科学进步的周期性本质这一首要论点。

首先,库恩针对科学家和科学共同体之间的关系提出了一个特别的观点。他把科学看成"社会企业",深受其从业者的特点以及"外部社会、经济和智力条件[1]"的影响。这代表了在科学哲学*研究方面一个不同寻常的发展。

在库恩看来,科学不同于人类的其他活动。在这一点上,他与赞同逻辑经验主义*思想学派的学者意见一致。但是逻辑经验主义者认为科学具有特殊性,因为它是知识十分理性化的积累。在这种情况下，库恩认为我们需反复才能获得知识，而不是仅仅被动地加工有关现实的数据。对库恩来说，科学家依据自己主动选择的范式，对"真"与否拥有话语权。

这一思想对 20 世纪 60 年代盛行的逻辑实证主义，特别是它对理性的假设是一个挑战，但也是一种冒险，有可能成为一种自我矛盾的论证。库恩提出，假如人类的努力既不是客观的，也不是人类知识独特的、线性的积累，那么，科学怎样在人类的努力中保持特殊地位呢？如果库恩试图说明科学活动和其他任何别的智力追求一样主观，那么这个矛盾就不会成为一个问题。

第二，库恩动摇了科学真理的概念。他关于范式的论点是，科学家不可能得出最佳结论，因为范式是由一个无法直接接触现实的科学共同体所形成和接受的，这就使得他在书末对"挽救'真理'概念[2]"做了评论。

第三，库恩探讨了科学的多元性。在他看来，不同的科学分支以它们自己的范式运作，有自己的周期性发展——这一概念摧毁了科学作为一种统一庞大企业的形象，其基本原则从不改变。正如库恩所看到的，科学探究的每一个分支——从生物学、化学和物理学到更专业的子学科——都有自己的共同体及范式；科学家和科学因研究领域和研究对象而异。

> 我认为，发达科学的研究者，无论是将其看作一个团体还是一些团体，根本上都是解谜者……解谜能力和其他价值一样，在应用中也会模棱两可。尽管如此，两个共享这一标准的人在使用时所做的判断也会不同。
>
> ——托马斯·库恩：《科学革命的结构》

思想探究

库恩把科学共同体视为一个封闭的单元,与社会分离,追求自己的议程。一旦科学家考虑并采用一种范式,他们便开始使用一种"深奥的"*语言——对普通大众来说难以理解的外来词汇[3]。

科学家们愿意选择追求那些有益于"人类福祉"的崇高目标[4];相反,库恩看到的却是他们的内在品质,他们只专注于改善和培养自己的声誉。库恩解释说,"与工程师、众多医生以及大多数神学家不一样,科学家无需去选择那些亟待解决的问题"[5];相反,"许多伟大的科学智慧者,都把自己的专业精力倾注在这种费解的谜题上。在绝大多数场合,任何特殊的领域除了解'谜'之外没有别的工作可做了,这一事实对于这种迷恋解谜的人来说更着迷[6]。"

通过这种方式,库恩推翻了原先对科学家的理解,即科学是一个为人类服务的共同体。它开启了科学领域,引起了科学社会学家的注意。他认为科学家和非科学家是一样的,以暗示他们的社会交往需要同样的审查[7]。然而,在提出这个论点时,他坚持认为科学本身与人类其他活动不同。

缺乏共识的协议——范式——即为空洞科学:"一旦发现了据以观察自然界的第一个范式,就再也不会有缺乏任何范式的研究工作了。拒斥一个范式而又不同时用另一个范式去取而代之,也就等于拒斥了科学本身[8]。"虽然后来社会科学家在别的领域里采用了这个概念,但这并不是库恩的目的。对他来说,范式是科学的决定性属性。

库恩对真理概念认识的不稳定性也与范式概念有关。在他看来,没有范式能够解释现实的真实面貌。最终,每一种范式都在衰退，让位于一种新的范式，在科学结论中没有任何永恒，只有创造持续的进步[9]。

这个概念衍生了相对主义思想，认为科学知识没有普遍有效性。相反，科学知识在任何特定的时候都与科学发展紧密相关。正如库恩写的那样，"对一个科学理论来说，极为成功绝不是完全的成功[10]。"在库恩的矛盾叙述中，范式越精确，失败的可能性就越大。在精确度提高的背景下，异常现象*显得更加明显。这种"技术故障"导致了常规科学*的危机[11]。所以我们"可能不得不抛弃这种不管是明确还是含糊的想法：范式的转变，使科学家和向他们学习的人越来越接近真理[12]。"

在常规科学时期，当研究工作中利用共识范式*产生大量发现时，就会产生根本进步。库恩认为，这是整个科学界达成共识的自然结果。这样，"对一个共同范式的接受，使科学界（摆脱）不断地重新审视其第一个范式的需求。"因此，其成员可以将注意力集中在"关注它的最微妙和最深奥的现象"上[13]。在这方面，进展可以定义为提高原始范式的精确性。

然而，令人吃惊的是，有一种感觉，跨范式的进步是可能的。正如库恩所看到的，一种范式从未完全消除先前范式的发现。相反，新范式"通常保留了过去成就中很多最具体的部分，此外还总能容许有额外的具体问题解决方法"[14]。含蓄地说，随着新发展与旧发展的结合，一种进步的感觉就出现了。正如库恩所言，科学发展如同企业，"以对自然日益细致和精确的理解[15]"为特征。

为了阐明这一观点，库恩以达尔文*选择为例，把英国博物学家查尔斯·达尔文*在19世纪的生物研究工作，从根本上描述为从远离目的论*（其思想是科学发展在一个预先决定的目标上进行）

的运动，到把自然界理解为只能以早期化身来衡量的意外进步。其结果是，提高了"表述的清晰程度和专业程度"，却没有明确的目标[16]。值得注意的是，库恩从来没有将这个比喻与达尔文进化论联系在一起，即使是在他的晚期作品中。

库恩的相对主义同科学家选择范式的过程紧密相连。一旦科学陷入危机，且相互矛盾的理论出现，范式的成功"绝不可能仅仅靠逻辑和实验明确地确定。[17]"相反，它会依赖以下因素：信仰、初高级科学家之间的权力斗争、大学行政管理对科学家个人的支持，以及可获得的研究资金和金融支持等。库恩在《科学革命的结构》中说得很清楚，范式不成功，因为它们是"真实的"，但最终因为科学界的主导派系选择了一种范式而不是另一种。

库恩提及小说《1984》，在英国作家乔治·奥威尔*所憧憬的未来里，个性被抹杀。库恩写道，科学共同体的新成员"就像……《1984》中的典型人物，是被权力重写的历史牺牲品。[18]"他或她接受的训练不是质疑范式，而是接受范式；从这个意义上讲，在科学界里教育基本上是一种"启蒙[19]"。

尽管如此，库恩还是在范式临时性和不同科学分支使用不同范式这一事实的基础上，提出了科学多样性的论点，试图形成支撑所有科学历史的基本概念。但他坚持认为，科学家有权利选择他们使用何种范式以及何时使用[20]。从这个意义上讲，范式的成功"绝不可能仅仅靠逻辑和实验明确确定"，这说明科学家的观点和偏爱在科学探索过程中起着作用[21]。库恩含蓄地拒绝了统一科学和万物理论的大概念，试图弄明白"科学中的真理是否可以成为一种理论[22]。科学革命发生在专门传统中，无需用于他人[23]。

在17世纪，科学家就我们认为理所当然的重力和运动定律做了很多发现，这无疑是一场伟大的科学革命，但是库恩认为科学革命并不需要如此大的范围。对于他来说，科学革命以不同的形式和规模发生，既包括著名的"革命"，如法国化学家安托万·拉瓦锡*发现氧的存在，也包括不太著名的 X 射线的发现[24]。

在库恩看来，这些革命有一个共同之点：它们都瓦解并取代了现行范式[25]。

被忽视之处

学者们经常忽视库恩在《结构》中提出的一些心理学和认知科学方面的观点[26]。

他对社区事务的社会分析，使格式塔*心理学中的"知觉变化"的观点黯然失色。英国科学史学家亚历山大·伯德解释说，较之社会学，库恩更懂心理学[27]："是该重新评价库恩思想中的自然主义元素的时候了。这些思想库恩本人都放弃了，如科学革命的心理学性质和不可通约性的心理学而不是语言学的理解[28]。"伯德特别提及认知科学*，把它当作这类分析的工具，但是他并未详细论述这一思想。

研究库恩著作中（被忽视的）认知概念的学者，包括科学哲学家汉纳·安德森*、彼得·巴克*和陈向[29]；所有这些都集中在库恩对哲学家路德维格·维特根斯坦*"家族相似性"理论较少讨论的支持上。维特根斯坦认为，看似由一个特征连接起来的物体，实则是由一系列重叠的特征将它们联系在一起[30]——这就是库恩范式概念背后的理念。根据安德森、巴克和陈向的介绍，库恩是英语世界里唯一的、在 20 世纪 70 年代中期之前接受这个思想并与之互动的科学哲学家[31]。

1. 托马斯·S.库恩:《科学革命的结构》第 4 版,伊利诺伊州芝加哥:芝加哥大学出版社,2012 年,第 X 页。
2. 库恩:《结构》,第 206 页。
3. 库恩:《结构》,第 20—21 页。
4. 库恩:《结构》,第 37 页。
5. 库恩:《结构》,第 163 页。
6. 库恩:《结构》,第 38 页。
7. 库恩:《结构》,第 40 页。
8. 库恩:《结构》,第 79 页。
9. 亚历山大·伯德:《托马斯·库恩》,切舍姆:欧酷曼出版社,2000 年,第 3—9 页。
10. 库恩:《结构》,第 68 页。
11. 库恩:《结构》,第 69 页。
12. 库恩:《结构》,第 169 页。
13. 库恩:《结构》,第 163 页。
14. 库恩:《结构》,第 168 页。
15. 库恩:《结构》,第 169 页。
16. 库恩:《结构》,第 171 页。
17. 库恩:《结构》,第 95 页。
18. 库恩:《结构》,第 166 页。
19. 库恩:《结构》,第 164 页。
20. 库恩:《结构》,第 50 页。
21. 库恩:《结构》,第 95 页。
22. 库恩:《结构》,第 167 页。
23. 库恩:《结构》,第 50—51 页。
24. 库恩:《结构》,第 93 页。
25. 库恩:《结构》,第 92 页。
26. 亚历山大·伯德:"科学革命的结构及其意义:五十周年纪念版论文评述",《英国科学哲学杂志》第 63 卷,2012 年第 4 期,第 865 页。
27. 伯德:《科学革命的结构》,第 7—8 页。
28. 伯德:《托马斯·库恩》,第 14 页。

29. 汉纳·安德森、彼得·巴克、陈向:《科学革命的认知结构》,剑桥:剑桥大学出版社,2006年,第12—18页。
30. 南希·内尔西塞安:"库恩、概念变化和认知科学",载托马斯·尼克尔斯编《托马斯·库恩》,剑桥:剑桥大学出版社,2003年,第180页。
31. 安德森、巴克和陈向:《认知结构》,第8页;内尔西塞安:"库恩和认知科学",第1页。

7 历史成就

要点

- 《科学革命的结构》实现了托马斯·库恩的目标,即从历史的角度重新审视自然科学。
- 科学和社会学*的外部*读物的增长(据此,政治和文化在科学进步中扮演重要角色),在很大程度上帮助库恩取得了巨大成功。
- 学术上对库恩作品的误解,妨碍了他向广大读者传递其观点的能力。

观点评价

托马斯·库恩《科学革命的结构》的主要目的是挑战这样的观点,即认为科学是一种不断增加人类知识量的理性*运动[1]。

尽管有人认为库恩也想要终结逻辑经验主义哲学*,但科学史学家迈克尔·弗里德曼*和乔治·A. 赖施*已经证明了他与经验主义者有着紧密的联系[2]。库恩的研究工作确实与德国逻辑经验主义者鲁道夫·卡尔纳普*的研究有惊人的相似之处[3]。因此,准确地说,尽管库恩的基本历史观点挑战了逻辑经验主义的历史和科学哲学*的学术主流,但他接受了这种"教科书传统"的一些核心理论。

库恩将自己视为最新获胜科学流派[4]的代言人,希望自己所选择的观点将把科学史*从假设中解放出来,这个假设就是科学史的作用只反映哲学。

揭示库恩这本书的真实目的是一件十分艰难的工作,并不是乍一看那么轻松；虽然他明确声明自己意在反驳"教科书传统"，但这一意图并不明显。这种模棱两可的态度导致评论家有各种各样的解释。在20世纪60年代革命时期，很多人愿意把库恩理解为不同凡响的、具有颠覆性的思想家。虽然他否认这种说法，但是他书中的许多段落似乎鼓励人们这样阅读。当这本书首次出版时，它与主流科学教科书的教条僵化形成了鲜明的对比[5]。

今天我们清楚地看到，库恩的著作使我们对科学的理解产生了巨大的变化。但是，我们很难将我们对这些变化的认识，与库恩在写这本书时的意图区分开来。虽然库恩对他那个时代有关科学的共同假设提出了挑战，但其本意不是挑战科学本身——这一点最终还是发生了[6]。

在挑战科学更古老的、累积的理解时，库恩当然就证明了科学是周期性的。不过，仍然很难说他完成了更多的工作。一部分困难源自库恩主要概念的模糊性（例如"范式*"）。此外，他的工作如此迅速、全面彻底地被社会科学家和科学史家所采纳，以致很难想像出有任何前库恩模型来理解科学的历史和发展。

> 库恩的书表达清晰，论述有力，充满着令人信服的例子，没有晦涩难懂的词汇和符号，否则对大多数非专业人士来说一窍不通。库恩的思想很快被远离自然科学领域的学者所接受。
>
> ——彼得·诺威克：《崇高梦想》

当时的成就

在20世纪中叶，通过关注科学研究发生的更大社会背景，大学增加了对科学"外部"阅读及其社会地位的探索。历史学家理查德·霍夫斯塔特*的著作《1860—1915美国思想中的社会达尔文主义》（1944）就是很经典的考察。霍夫斯塔特探讨了"适者生存""生存竞争"等词组，是怎样以意想不到的方式从查尔斯·达尔文*的进化生物中出现，进入社会评论，再反馈到科学实践[7]。这种趋势当然有助于《科学革命的结构》的成功，所以这本书一出版就激发了人们广泛的兴趣而大受欢迎。

就此而言，不得不提及社会学——这是一个特别关注科学和思想历史性地受社会条件影响而形成的领域。例如，科学哲学家詹姆斯·马科姆*认为，没有哪一门学科比社会学更能支持库恩的著作，尤其是因为库恩将科学团体*看作一个社会群体[8]，在这个团体中寻找科学的原因和特征。

此外，《科学革命的结构》的出现，就像第二次世界大战*中使用原子弹的公开辩论那样，变得更加广泛。库恩青年时代就受到了这个问题的影响，他在冷战*高峰时期写了《结构》，当时任何人都会从社会或外部主义角度批判科学；许多持有（或者怀疑持有）这种观点的教授都失去了工作[9]。

库恩代表了新一代学者，他们愿意改变科学现状，并质疑主流的科学观点，认为这是客观无疑的。辩论的时机已成熟，正如英国科技翻译齐亚乌丁·萨达尔*和加拿大科学哲学家伊安·哈金*指出的那样，20世纪60年代，公众和知识界开始质疑诸如库恩早期导师詹姆斯·布赖恩特·科南特*等科学家在制造原子武器中所起的作用[10]。从这个意义上来讲，库恩的著作标志着我们看待社会科学的方式发生了深刻的转变。

《结构》这本书对科学新观点的产生起到了一个跳板的作用。它是新学术和知识运动发展的一个重要里程碑，使科学拥有科学家的社会与知识背景的功能[11]。这些思想家以全新的方式质疑了现代科学，强调现代科学是一种具体地理区域（西欧）的产物、一种特殊的文明（犹太教和基督教共有的）和一种具体性别（男性）。有的甚至质疑了科学对真理和知识的主张。

这些观点让库恩感到吃惊，有的还让他感到忧虑。他从来没有想到他的著作有这样的潜力，能够广泛引发对科学和非科学主题进行重新评价，如知识社会学。他的著作所提供的分析工具，用途要比他所预期的要广泛得多。

正如萨达尔指出的，《结构》主要对科学内部变革感兴趣。库恩支持科学从社会中获得自主。在随后进行的处置中，他"纠正"了由他的书引起的错误观念，其目的是使科学免受公众关注[12]。事实上，美国哲学家史蒂夫·福勒*指责《结构》促进了保守形式的学术研究，并对科学的特殊本质产生了长期的偏见。福勒认为，有些人利用这个文本使激进的科学哲学边缘化。有两个例子，是学者杰瑞·莱文兹*和保罗·费耶阿本德*提出的哲学思想，他们都认为科学完全是由科学家个体的嗜好和偏好引起的，其行动与科学共同体*组成部分无关，也无产生关联的欲望[13]。

局限性

尽管《结构》一书影响持久，倍受欢迎，但其局限性仍然值得思考。其中最重要的是，这部著作是对科学特殊地位的维护，而这是20世纪历届政府和社会赋予的——这是一种基于假设的维护，这些假设是库恩从被称为逻辑经验主义*的哲学流派继承而来的。

然而，假如库恩是对的，科学知识并不完全是客观的，只是创造知识团体的产物，那么它是如何值得拥有并保持这种特殊地位的?撇开这一点,科学往往受限于大学有限资金支持的财政压力和奖金的诱惑。那么,科学研究将如何保持其严密性呢?制药公司为那些能研制出新的药物化合物的科学家提供了高额酬金。这是否偏离了科学界的研究议程?库恩并没有考虑这其中的奥秘。

此外,库恩只对科学界如何塑造科学实践的方式进行了初步探索。但是,面对动荡的现代经济,科学界内部的社会结构可能脆弱不堪,库恩对此并没有讨论——这正如在1973年的石油危机*中所暴露的那样,当时石油价格上涨引起了经济的不稳定。例如,大学在日益稀缺的资源方面面临着艰难的选择,这影响了学术环境的社会层面。然而,尽管科学界的内部文化并不像库恩所设想的那样稳定,它会随着时间和地域的不同而变化[14],但读者发现这一因素并没有包含在库恩的描述之中。结果对其范式和不可通约性*概念的历史分析仍然不完整,因为科学曾经享有的特殊地位渐渐失去[15]。

---

1. 托马斯·S. 库恩:《科学革命的结构》第4版,伊利诺伊州芝加哥:芝加哥大学出版社,2012年,第X页。
2. 迈克尔·弗里德曼:"库恩与逻辑经验主义",载托马斯·尼科尔斯编《托马斯·库恩》,剑桥:剑桥大学出版社,2003年,第19—21页;参见乔治·A. 赖施:"库恩扼杀了逻辑经验主义吗?",载《科学哲学》1991年第58卷,第266—267页。
3. 古洛尔·厄兹克和泰奥·格伦伯格:"卡尔拉普和库恩:死敌还是盟友?",《英国科学哲学杂志》第46卷,1995年第3期,第305页。
4. 库恩:《结构》,第135—138页。
5. 库恩:《结构》,第166页。

6. 彼得·格弗里-斯密斯:《科学哲学导论:理论和现实》,伊利诺伊州芝加哥:芝加哥大学出版社,2003年,第99页。
7. 理查德·霍夫斯塔特:《1860—1915美国思想中的社会达尔文主义》,波士顿:比肯出版公司,1944年,第6页。
8. 詹姆斯·A.马科姆:《托马斯·库恩的革命:科学的历史哲学》,伦敦:连续出版公司,2005年,第142页。
9. 齐亚乌丁·萨达尔:"托马斯·库恩和科学战争",载理查德·阿皮尼亚内西编《后现代主义和大科学》,剑桥:艾肯图书公司,2002,第197页。
10. 萨达尔:"托马斯·库恩和科学战争",第195页;哈金:"导读",载库恩:《结构》,第IX页。
11. 萨达尔:"托马斯·库恩和科学战争",载《后现代主义和大科学》,第211—221页。
12. 萨达尔:"托马斯·库恩和科学战争",第221—224页。
13. 史蒂夫·福勒:《托马斯·库恩:当代哲学史》,伊利诺伊州芝加哥:芝加哥大学出版社,2000年,第212页。
14. 约翰·M.齐曼:《真正的科学:它是什么,它意味着什么》,剑桥:剑桥大学出版社,2000年,第4—5页。
15. 齐曼:《真正的科学》,第1页。

8 著作地位

要点

- 托马斯·库恩的著作关注的是科学实践的历史解释,对科学实践的基本原则进行了深刻的社会学*和哲学思索。
- 库恩的第二部著作《科学革命的结构》,为他后来的研究规划了路线。
- 库恩最著名的、引起最广泛辩论的和最受批评的著作,仍属《科学革命的结构》。

定位

托马斯·库恩花了15年时间撰写他的第二部著作《科学革命的结构》[1]。他的第一部著作《哥白尼革命》(1957)源于他对经典力学*(即20世纪之前就存在的有关运动和力量的物理定律)的研究,但缺乏《结构》一书所展现的创新性。

1956年,哈佛大学终身教授委员会在库恩的资格申请表上,评阅了其对尚未出版的《哥白尼革命》的描述。认为这部著作与其说是对科学知识的贡献,不如说是一种推广科学的尝试[2]。他没有获得终身教职(一个永久学术职位的担保)。不过,由于这本书包含了一些早期版本的思想,这些思想成为《科学革命的结构》和库恩后期著作的核心,所以还是让研究库恩的学者感兴趣。

16世纪,波兰天文学家尼古拉·哥白尼*创立了理论,即是太阳而不是地球位于宇宙的中心。几个世纪以来,人们曾认为,世界观的成功靠的是哥白尼在天文学领域里独特的见解。然而,库恩认为，哥白尼写作年代的社会和学术氛围也起着重要作用³。在研究他的书时，库恩能够对革命思想做一个案例研究——这种思维模式将在《科学革命的结构》中成为"非常规科学*"的核心。

库恩的《科学革命的结构》中，科学发展规划的第二个主要组成部分与常规科学*有关。库恩首先在《现代物理科学中测量的功能》（1961）这篇文章中表达了这个思想。在此文里，他提出，把常规科学看成是将科学革命*的成果应用于日常科学实践的过程，认为"测量的功能及其特殊吸引力很大程度源自神话⁴。"

库恩早在《必要的张力》中就已经介绍了范式*的概念，这是他于1959年提交给犹他大学研究会议的文章⁵。

> 在托马斯·S.库恩的《科学革命的结构》里，有一种主要概念，时而被称为"历史主义"，时而被称为"相对主义"。虽然库恩教授一再强调，对他观点的解释，大多数都歪曲了自己的意思；但是目前还不完全清楚，他是否已经成功地回答了那些对其著作做出这样解释的批评者，也不清楚他是否澄清了自己的立场，以便这件事不再公开辩论。
>
> —— 莫里斯·曼德尔鲍姆："关于托马斯·S.库恩的《科学革命的结构》的说明"

整合

尽管他后来的著作颇丰，但《科学革命的结构》，尤其是1962年的第一版，定义了今天的库恩，塑造了一位影响持久的、有远见卓识的、激进的作者形象。知识界及大众继续助推这一神话。但是，虽然《结构》第一版中的意象和修辞确实显得有些激进，但最近的学术研究表明，库恩后来修改、删除并低调处理了这部著作中的许多主张和含义。

总的来说，在他出版的著作中，库恩保持着同样的意图：从历史的角度，同时参考科学共同体*重新评估科学如何改变和发展。他始终强调科学中的非科学层面和科学发展的历史动力，这一点从未改变。他也没有对范式和不可通约性*的观点做出承诺，尽管这受到了严厉和持续的批评。但是，《科学革命的结构》第一版中看似激进的语言与库恩后来的作品形成了鲜明的对比[6]。

学者们仍在争论这一点：在与科学哲学家发生冲突后，库恩是否放弃了激进的立场？或者，他的修辞中的革命语气夸大了他的意图？

意义

《科学革命的结构》让库恩名声大噪。这本书很快成为极有影响的文献，不仅对历史和科学哲学*，对其他学科也是如此。虽然，库恩在1970年后把自己大部分职业时光都花在澄清他认为评论家对其误解的方面[7]，但他仍然相信《科学革命的结构》是自己的得意之作[8]。

在《结构》之后，库恩转向纯科学史研究*，并一直坚持到20世纪70年代末。《黑体理论和量子不连续性，1894—1912》（1978）是一种直截了当的历史分析，没有提到《结构》这本书的思想[9]。他讨论了20世纪初德国理论物理学家*马克斯·普朗克*的部分论著。长期以来，评论家们一直认为，普朗克1900年和1901年的量子物理学*论文已经催化了经典到量子力学*的转变。然而，库恩采取相反的观点，认为普朗克实际上从来没有回避经典力学的世界观[10]。

事后看来，库恩认为，《黑体理论》是他在一个特定历史学科上的最佳作品。然而，历史学家、哲学家和物理学家并不这样认为，他们有些人甚至埋怨库恩并没有使用自己在《科学革命的结构》中设立的理论框架[11]。为了回应他们的批评，库恩在此书修订版中加进了一个后记，试图阐明与《结构》中思想的关系[12]。

1978年之后，库恩再次回到不可通约性这个概念上来，这一次是通过语言学的框架来研究的。为了做到这一点，他尽量让自己置身于格式塔心理学*及其感知理论之外。根据格式塔心理学，个体在基于信念的理论框架上构建自己的感知——即如何看待世界，如何对世界做出反应，以及如何同世界互动。库恩认为，人们在他的理论和格式塔心理学之间所建立的联系，导致他们误解了他的不可通约观。太多的学者觉得，他认为科学家只是在构建竞争性的范式而不涉及现实。在这种情况下，范例的不可通约性似乎仅仅来自人们的论点。

库恩想彻底转变了学者们对自己著作的误解，所以他开始针对《结构》引起困惑的部分开设讲座。例如，虽然他从未就这个话题出版过著作，但是他仍然就现实的科学知识与语言结构之间的关系开设讲座[13]。他的理论是，科学范式与特定于该范式概念和术语的词汇相关联，但是词汇也专门针对科学领域及其运作的科学共同体。所以研究不同领域使用不同范式的科学家，不能借鉴其他范式来互相帮助，而所有范例的概念框架都是独一无二的。这就是为什么在"科学"中有如此多的多元性——在现实中，单一的科学是由子学科和拥有各自范式的科学家团体组成的集合[14]。

1977年，库恩进一步扩展了他对"不可通约性"论题的思想，详细讨论了科学家如何选择一种范式作为共识的方法——他称之为"理论选择"的过程。他认为，在意识到《结构》的缺点后，这个概念已经成熟了，可以进一步阐明。特别是，他没有回答这个问题："如果科学家最终因为种种原因选择了一种共识范式*，而不是纯粹的科学考虑，那么他们的选择是否会是非理性的呢？"

库恩想要避免非理性的暗示*，因为它与他把科学作为一种自主、特殊活动的意图相冲突[15]，便重申并阐述了五种科学考虑，这五点对科学界的人们有指导作用，他们有能力和权威来决定一个共识范式：

- 精确性
- 一致性
- 范围的广度
- 简洁性
- 成效性[16]

因此，尽管科学界的社会学特征决定了哪些科学家有能力选择共识范式，科学因素的考虑形成了他们最终的选择。

---

1. 托马斯·S.库恩：《科学革命的结构》第 4 版，伊利诺伊州芝加哥：芝加哥大学出版社，2012 年，第 XXXIX 页。
2. 詹姆斯·A.马科姆：《托马斯·库恩的革命：科学的历史哲学》，伦敦：连续出版公司，2005 年，第 13—14 页。
3. 托马斯·S.库恩：《哥白尼革命：西方思想发展中的行星天文学》，坎布里奇：哈佛大学出版社，1957 年，第 VIII 页。

4. 托马斯·S.库恩:"现代物理科学中测量的功能",载《必要的张力:科学传统和变革研究文选》,伊利诺伊州芝加哥:芝加哥大学出版社,1979年,第161页。
5. 托马斯·S.库恩:"必要的张力",载《必要的张力》,第18—26页。
6. 彼得·诺威克:《崇高梦想:"客观问题"和美国历史行业》,剑桥:剑桥大学出版社,1988年,第526—528页。
7. 托马斯·S.库恩:《结构以来的路:1970—1993哲学论文集与自传式访谈》,詹姆斯·科南特和约翰·海于格兰编,伊利诺伊州芝加哥:芝加哥大学出版社,2000年,第7页。
8. 马科姆:《托马斯·库恩的革命》,第109页。
9. 托马斯·S.库恩:《黑体理论和量子不连续性,1894—1912》,伊利诺伊州芝加哥:芝加哥大学出版社,1978年。
10. 马科姆在《托马斯·库恩的革命》这部著作中简短而有益的总结,第108—112页。
11. 马科姆:《托马斯·库恩的革命》,第109页。
12. 库恩:《黑体理论》第2版,1987年,第349—379页。
13. 马科姆:《托马斯·库恩的革命》,第24页。
14. 霍华德·桑克:"库恩不断变化的不可通约性概念",《英国科学哲学杂志》第44卷,1993年第4期,第770—772页,描述了库恩不可通约性理论中的变化。
15. 库恩:"客观性、价值判断与理论选择",载《必要的张力》,第321—322页。
16. 库恩:"客观性、价值判断与理论选择",第320—339页。

第三部分：学术影响

9 最初反响

要点

- 批评者最初关注库恩的范式*和不可通约性*等概念、他们在其著作中所感受到的相对主义*（即大致上说，"完美"知识是不可获得的意思），以及他所展示的科学界形象。
- 最有影响的批评来自哲学家达德利·夏皮尔*和卡尔·波普尔*，他们抓住库恩的范式概念缺乏精确性的问题，波普尔还特别抨击了库恩把科学进步理解为周期性的观点。
- 早期的许多反应都受到科学的政治文化地位的影响，这是冷战*时期自由社会的一部分。

批评

在出版的头两年里，托马斯·库恩的《科学革命的结构》还是获得了积极的评论，但在1964年之后，批评的声音开始出现了，主要来自科学哲学家。批评者集中在三个方面，关注"范式"一词（通过联想，《结构》描绘了科学和科学共同体的方式）、他们所认为的文本相对主义和理想主义*。

1964年，科学哲学家达德利·夏皮尔批评库恩的范式定义，认为这个概念太不确切[1]。他还不赞成库恩的观点，即范式转移*可以改变"质量"等科学概念的实际意义。相反，他认为它可能会改变这种观念的应用[2]。最后，夏皮尔发现不可通约性和范式的概念是无法调和的。他指出，理解不可通约性意味着完全的"世界变化"，范式之间并不能真正产生分歧[3]。

影响很大的奥地利裔英国籍哲学家卡尔·波普尔也批评了库恩，认为其术语"范式"缺乏精确性。在1965年伦敦经济学院的一次论坛上，波普尔及其追随者与库恩进行辩论[4]。支持波普尔的科学哲学家批评库恩，其中包括库恩的伯克利同事保罗·费耶阿本德，但是很难详细知道这场争论对库恩的名声影响有多大。因为，正如一位评论家所观察到的，在1965年，"库恩和波普尔对科学的看法，可能更多地是因声誉而不是因读者量为人所知[5]"，所以这一场讨论可能永远不会仅仅包括一小群学者。在波普尔看来，科学在本质上具有革命性的特性，其基本特征包括"大胆猜想"、不断地重新检验和驳斥其理论[6]。研讨会上，波普尔和其他学者完全不同意库恩把科学区分为"常规的"和"非常规的"。对他们来说，正如库恩的翻译和拥护者保罗·霍伊宁根-休内*指出的，科学是一种"由人类认知能力不可靠性的持续意识塑造的"企业[7]。

波普尔继续强烈批评库恩常规科学*的思想，认为是一种"对科学，甚至对我们文明的危险"[8]。在波普尔的观念里，库恩思想里的教条特征是极为不科学的[9]。虽然波普尔承认存在教条化的科学事业，但是他把"常规科学"看成自相矛盾的术语[10]。波普尔的追随者对库恩有关科学共同体的描述也不满意。他们相信科学完全是开放的，严厉批评库恩把科学描述成"一种封闭的社团，其主要特征是'摒弃评判性话语'"[11]。

最强烈的批评是科学哲学家们提出来的，他们认为库恩支持了非理性*和相对主义。正像著名学者、库恩著作的译者和编辑瓦索·金迪*和西奥多·阿拉贝齐*最近所指出的那样，这些哲学家担心库恩的观点过于泛化。库恩曾经说过，在科学知识范围之外的考虑（例如，科学界的专业竞争和权力斗争，以及科学家的个人观点)决定了科学知识的方法和结论。许多哲学家担心这样会偏离太远[12]。他们不喜欢库恩的基本观点,因为它违背了科学主张,在一个毫无疑问的客观领域里,站在学术争论之上。

像波普尔所看到的那样,库恩认为,科学家不能理性地决定采用哪种框架[13]。匈牙利科学哲学家伊姆雷·拉卡托斯*指责库恩提出了一种以"群众心理"为特征的科学变化观,还公开谴责了库恩的范式变化观点是一种"神秘的转换,它不是也不能被理性规则所支配"[14]。

> "看,"托马斯·库恩说。这个词说得很沉重,带着疲倦,好像他被迫接受被我误解的事实,但他还是要试一试——这毫无疑问是徒劳的——为了阐明他的观点。"看,"他又说。他把瘦长的身躯和闷闷不乐的脸向前倾斜着,他那宽厚的下唇,常常和蔼地蜷曲在嘴角,下垂着。"看在耶稣基督的份上,如果我可以选择写这本书或者不写这本书,我就会选择写,但肯定在某些方面对它的回应感到相当不安。"
>
> ——约翰·霍根:《科学的终结》

回应

库恩对同侪的批评非常敏感,便在伦敦的一些学术会议上(1965)、斯沃斯莫尔讲座上(1967年也是在伦敦)和1969年在乌尔班纳举行的学术会议上宣读了系列论文[15]。他坚持认为自己并没有提倡相对主义和非理性。在长期的回应中,他断言科学"是我们完备知识的最可靠的例子"。虽然没有依据传统视角,但是他还是表达了自己对科学进步的信念[16]。

在《科学革命的结构》中,他试图以进化论为类比来阐明自己的思想。查尔斯·达尔文*提出了适者生存的理论。如同达尔文的进化论,科学的发展没有明确的目标,只朝着一个方向不可逆转地前进。库恩特别不能接受这一观点:科学发展使知识更加接近真理[17]。

除此以外,库恩还在描述意义和规范意义*(即告诉读者科学应该如何,而不是它实际如何)上强烈维护了常规科学的概念。他声称,只有在常规科学的背景下才能发现革命科学*。他还坚持认为,常规科学的静止状态有助于深化科学研究促使其进步。在1963年出版的引发广泛争论的论文中,在1965年与卡尔·波普尔及其支持者的辩论之前,库恩就"教条"对科学研究的作用与重要性进行了辩护。他后来之所以放弃使用这个词,是因为它有负面意义[18]。

库恩还特别努力澄清了范式概念——批评者认为这是一种既不清晰又完全不连贯的概念。他承认,他起先使用这个术语确实太模糊。《科学革命的结构》中,"范式"这个术语的"过度塑性"导致把这个概念不恰当地使用在其他学科中,特别是社会学中[19]。他试图把它分成"广义"和"狭义"两种含义,将广义比作一个学科基质*(一个科学团体所使用的符号概括、模型和问题解决方案[20]),将狭义用于解决具体问题。

德国科学哲学家保罗·霍伊宁根-休内观察到,库恩在他后期著作中不再使用学科基质这个术语,因为他关注了范例*这个概念——意思是,在常规科学时期,科学家对特定问题的解决方案在共识范式*下,成为众所周知的行为模型(即"范例")*[21]。另一个变化是,库恩意识到"范式"既可以指普遍的共识,也可以指某些学派思想的内部共识。

库恩还想澄清不可通约性概念,以期减少对这个概念的激进理解。在《结构》的初版里,库恩把不可通约性多少有些模棱两可地解释为"世界变革"。这使读者以为,库恩打算用范式转移来暗示完全取代概念框架[22]。事实上,库恩后来声明,他从来无意改变整个概念框架——只是调整一些理论、术语、词汇或语言[23]。因此,沟通仍然是可能的,不过只是部分的。

这个方式让库恩进一步强调语言和翻译问题[24]。在这一点上,他提到了科学哲学家蒯因*的著作。蒯因的模糊语言理论认为,翻译无法做到完美[25]。库恩利用蒯因的观点来说明,跨范式之间所有术语的翻译都是不完整的。但同时,也有些术语含义相同,这就是交流可以继续下去的地方。

库恩希望,这些修改能够阐明新理论的选择并非完全不理性;他认为有些方面仍然适合用来比较。其中,包括关于特定情况的经验*预测和断言(即基于可观察的证据和理性推理的预测和断言),以及两个相互矛盾理论的比较[26]。他强烈否定不可通约性和不可比较性是同一回事[27]。

冲突与共识

尽管库恩对《科学革命的结构》的修改极大地改变了他的著作,但并没有说服所有的批评者。他们中的许多人认为,库恩的澄清是一种新的、不那么激进的理论[28]。

科学哲学家约翰·沃勒尔*曾在伊姆雷·拉卡托斯指导下做研究,批评了库恩在进步这个话题上的退让。沃勒尔认为,淡化不可通约性的概念,会贬低他的整体论点[29]。然而,库恩仍然坚定地认为，他的解释是为了澄清——而不是修改——他的想法。

有些批评者并没有完全接受1970年出版的《科学革命的结构》的修订版。例如，科学哲学家艾兰·马斯格雷夫*和达德利·夏皮尔，对库恩使用学科基质和范例来解释范式并不信服[30]。按理说，夏皮尔是库恩最激烈的批评者，也感觉到库恩并没有成功地为自己辩解，反驳别人谴责相对主义[31]。根本问题源自《结构》的第一版，正如夏皮尔所言，书中库恩的一些关键概念，如范式，仍然晦涩难懂。一个人不能通过进一步解释荒谬的概念来产生一个可理解的概念[32]。

围绕《科学革命的结构》的争议并没有很快消失。虽然在随后的几年里不是那么火热，但是库恩继续成为批评的对象，特别是来自科学哲学家的批评。库恩甚至激怒了自己的支持者，特别是那些理性的学者和思想家，对他后期收回观点并不满意。

尽管如此，他仍然坚持认为，写《结构》这本书是对科学史和科学哲学*的及时贡献。在接受采访时，他评论道："如果我可以选择写这本书或者不写这本书，我愿意会选择写这本书，但这样肯定在某些方面会对它的反应感到相当不安。[33]"

---

1. 达德利·夏皮尔："科学革命的结构"，《哲学评论》第73卷，1964年第3期，第388页。
2. 夏皮尔："科学革命的结构"，第390页。
3. 夏皮尔："科学革命的结构"，第391页。
4. 库恩和夏皮尔之间冲突的故事，在史蒂夫·福勒编纂的《库恩对夏皮尔：科学灵魂之争》(纽约:哥伦比亚大学出版社,2004年)当中有很好的描述。

5. 福勒:《库恩对夏皮尔》,第29页。
6. 卡尔·波普尔:"常规科学及其危害",载伊姆雷·拉卡托斯和A.马斯格雷夫编《批判与知识的增长》,剑桥:剑桥大学出版社,1975年,第55页。
7. 保罗·霍伊宁根-休内:《重构科学革命:托马斯·S.库恩的科学哲学》,伊利诺伊州芝加哥:芝加哥大学出版社,1993年,168页。
8. 波普尔:"常规科学",第53页。
9. 詹姆斯·A.马科姆:《托马斯·库恩的革命:科学的历史哲学》,伦敦:连续出版公司,2005年,第86—87页。
10. 约翰·沃勒尔:"常规科学与教条主义、范式与进步:库恩'对'波普尔和拉卡托斯",载托马斯·尼科尔斯编《托马斯·库恩》,剑桥:剑桥大学出版社,2003年,第67—69页。
11. J.沃特金斯:"反对'常规科学'",载伊姆雷·拉卡托斯和A.马斯格雷夫编《批判与知识的增长》,剑桥:剑桥大学出版社,1975年,第37页。
12. 瓦索·金迪和西奥多·阿拉贝齐:"导论",载瓦索·金迪和西奥多·阿拉贝齐编《重读库恩的科学革命的结构》,纽约:劳德里奇出版社,2012年,第2页。
13. 马科姆:《托马斯·库恩的革命》,第87页。
14. 伊姆雷·拉卡托斯:"证伪与科学研究纲领方法论",载伊姆雷·拉卡托斯和A.马斯格雷夫编《批判与知识的增长》,剑桥:剑桥大学出版社,1975年,第87页。
15. 马科姆:《托马斯·库恩的革命》,第101页。
16. 托马斯·S.库恩:"发现的逻辑学还是研究的心理学?",载拉卡托斯和马斯格雷夫编《批判与知识的增长》,第20页。
17. 库恩:"发现的逻辑学",载拉卡托斯和马斯格雷夫编《批判与知识的增长》,第1页。
18. 托马斯·S.库恩:"科学研究中教条的作用",载阿利斯泰尔·C.克龙比编《科学变革》,纽约:基础图书出版公司,1963年,第347—369页。
19. 托马斯·库恩:"关于范式的再思考",载《必要的张力:科学传统和变革研究文选》,伊利诺伊州芝加哥:芝加哥大学出版社,1977年,第259、295—319页。
20. 保罗·霍伊宁根-休内在《重构科学革命》中对学科基质成分的详细分析,第145—159页。
21. 保罗·霍伊宁根-休内:《重构科学革命》,第143页。

22. W.H. 牛顿-史密斯:《科学的理性》,伦敦:劳德里奇出版社,1981年,第12页;希拉里·普特南:《理性、真理和历史》,剑桥:剑桥大学出版社,1981年,第115页。
23. 托马斯·库恩:"后记",载于《科学革命的结构》第4版,伊利诺伊州芝加哥:芝加哥大学出版社,2012年,第197—203页。
24. 霍华德·桑克:"库恩不断变化的不可通约性概念",《英国科学哲学杂志》第44卷,1993年第4期,第765页。
25. 库恩:"后记",载《结构》,第201—202页。
26. 保罗·霍伊宁根-休内:《重构科学革命》,第219—221页。
27. 托马斯·S.库恩:"作为结构变化的理论变化:评斯尼德形式主义",《认知》第10卷,1976年第2期,第191页。
28. 牛顿-史密斯:《科学的理性》,第113—114页;普特南:《理性、真理和历史》,第126页;M.V.柯德:"库恩、科学革命与哥白尼革命",《自然与系统》第6卷,1984年,第4页。
29. 沃勒尔:"常规科学",第93页。
30. 马斯格雷夫:"库恩之二次思考",第293页;夏皮尔:"范式概念",《科学》第172卷,1971年,第707页。
31. 夏皮尔:"范式概念",第708页。
32. 夏皮尔:"范式概念",第710页。
33. 彼得·格弗里-斯密斯:《科学哲学导论:理论和现实》,伊利诺伊州芝加哥:芝加哥大学出版社,2003年,第87页。

10 后续争议

要点

- 《科学革命的结构》成为科学社会学家的参考点;而在职科学家自己作为一个不断扩大的学者群体,也在争论科学概念,而且经常立足于各自的背景争论。

- 虽然当今还没有库恩学派的思想,但他的著作在哲学、历史、社会学*、心理学和自然科学领域都产生了广泛的影响。

- 库恩的著作在科学知识的社会学分支学科*中引发了一场新的辩论,探究科学家们的品味如何影响他们的结论。

应用与问题

托马斯·库恩的《科学革命的结构》被认为是影响深远的著作,对研究科学史*的学者及学习社会学和知识哲学(以及更广义的社会学)的学生如此,对在教育和非营利性部门从事促进大众参与科学的人来说亦如此。

和他同时代的哲学家保罗·费耶阿本德*、卡尔·波普尔*和诺伍·德罗素·汉森*一样,库恩对科学进行了彻底的重释,结束了逻辑经验主义*学派的统治。做得比同时代的人要多的是,"库恩描绘了一幅非常生动的科学变革蓝图,改变了科学哲学*。[1]"

虽然《结构》在第一次出版时引起了相当大的争议,但也许很难发现它所提出的思想和问题在后来学者的著作中造成的影响。

科学哲学家伊安·哈金*于1981年提出一个观点,认为《科学革命的结构》终结了几个主要的概念。可以说,这其中最引人注目的是现实主义*的概念——即科学可以发现关于现实世界真理的观点[2]。其他学者则认为《结构》对科学哲学影响极小。例如，分别是丹麦、英国和美国的科学哲学家汉纳·安德森*、彼得·巴克*、陈向*各自都认为，科学哲学家对库恩的著作之所以产生了负面的看法，是因为他们的错误已经持续了近半个世纪。他们认为，当代科学哲学脱离了历史学研究，反而集中于维护现实主义立场（即科学事实只能通过观察来决定），反对建构主义*立场（即科学家通过观察决定事实，再根据自己的信仰解释事实）[3]。

> 科学哲学家蒂姆·莫德林雄辩地指出，最基本的问题是，存在两个库恩——一个温和的库恩和他的不温和的兄弟——在《科学革命的结构》整本书中相互推挤。
> —— 艾伦·索卡尔和珍·布里克蒙特:《学术欺诈行为》

思想流派

库恩的作品介于历史思想潮流和科学哲学之间，没有严格的库恩学派存在。学者们一致认为，库恩的著作被证明对科学史学家具有高度的争议性，即使这些史学家很少采纳库恩的主张。例如，美国科学史家简·格林斯基*声称，库恩的"在历史学家当中的影响是极其有限的[4]"（学者瓦索·金迪*和西奥多·阿拉贝齐*也持这一观点）[5]。

尽管库恩的学生努力在科学史界传播老师的学术方法，但是这一观点持续存在。例如，美国院士约翰·L. 海尔布龙*写道，库恩"让我们明白，我们正在进行一场伟大时代的智力冒险[6]。"科学历史学家——特别是 N. M. 斯维尔德洛*、J. Z. 布赫瓦尔德*和诺顿·外斯*等人——都在采用库恩的科学共同体*、科学革命*和不可通约性*等概念，来推进他的研究议程和对科学进步的周期性理解[7]。

20世纪70年代，去库恩著作化的运动催生了科学家进行研究的外在论*方法。这些方法强调，科学是外部因素的产物，如社会和政治事件，却对科学准确性和完整性的要求关注得较少，即使有的话。

20世纪90年代，这一领域又出现了新的问题，库恩对此没有答案。库恩批判了以前对科学的理解，具有讽刺意味的是，这样产生了一个学科，驳斥了把他看作是"过时的宏大叙事风格"[8]的代表。换言之，历史学家选择放弃一些内在论*的假设，比如库恩在解释科学家如何选择共识范式*时强调的假设，转而采用新的方法，尤其是社会学的方法，试图传达不仅仅是从内部看到的关于科学的"苍白的*陈词滥调"[9]。

这种运动一直持续到20世纪初和21世纪初，随着外在论者阅读浪潮的兴起，人们关注的是有影响力的法国哲学家米歇尔·福柯*的著作。福柯认为，社会决定因素在寻求理解过程中是至为重要的[10]。

当代研究

目前，激进的知识分子仍然是托马斯·库恩论著最热心的支持者。《科学革命的结构》已经成为后现代主义*科学评论的基础文献，质疑了科学客观性的可能性，认为它是一种形式的文化产物，就像其他任何事物一样，受到语境和解释等相同问题的影响。这些方法激发了诸如女权主义*和后殖民*研究等新的学术分支。这些新的学科试图证明，妇女和殖民题材的研究，虽然被之前的学者所忽视，但是在科学史上仍有一席之地[11]。正如科学翻译齐亚丁·萨达尔*指出的那样，库恩被认为是"颠覆科学"的人，他们渴望保留逻辑经验主义者所呈现不多的科学内在论文献（指的是这些人，他们试图维护科学家主张的客观性，同时探索个人和职业因素是如何影响科学实践的）[12]。

库恩的思想仍然是科学知识社会学的核心[13]，这是由爱丁堡大学的社会学家大卫·布鲁尔*和巴里·巴尼斯*提出的方法，坚持认为社会条件创造了科学知识[14]。法国社会学家布鲁诺·拉图尔*和他的同事斯蒂芬·沃格*举出了这个方法最为蹩脚的例子。"我们最普遍的目标"，他们在《实验室的生活》（1969）的前言中写道，"是为了阐明'科学软肋'的本质：因此，我们关注的是一位科学家牢牢地在他的实验室工作台上所做的工作。[15]"他们所说的"软肋"指的是可塑的条件。他们声称，这种条件形成了科学。

最近，后现代主义科学批评家把关注点转移到了科学对社会的作用上。如萨达尔所观察的，科学是一个有相当大权力和权威的领域，是一个把全球化*的法人和政府资本紧密联系在一起的领域。这些最新的批评家把他们的辩论集中在对科学公众监督的重要性和水平上。

科学的激进批评家认为，今天科学与政治经济权力结构——即政府与跨国公司——联系得如此紧密，以致历史学家和科学哲学家作为中立的或公正的观察者，需要对科学家的活动进行仔细审查，就像说客的动机和利益冲突以同样的方式接受审查一样。这些批评家也对西方国家每年在科学研究上花费数十亿提出了质疑[16]。

一些人仍然支持科学应独立自主的观点。这些批评家反驳道，

"学术左派"不能恰当地理解科学是什么、它是如何运作的,其内在实质是对多元文化主义*语言的批判。数学家、物理学家艾伦·索卡尔*曾经说过,在一个高度重视多元文化的环境中,"不可理解性成为一种美德;暗指、隐喻和双关语代替了证据和逻辑[17]。"他认为,后现代主义学术研究导致科学的公共诚信缺失——在他看来,缺失妨碍进步。生物学家 E. O. 威尔森在 1994 年的一次演讲中简明扼要地(并且冷嘲热讽)说到,多元文化主义等于相对主义*,也等于没有超级对撞机*[18]。换言之,索卡尔和威尔森都认为,一旦科学著作的诚信出了问题,这正如自库恩以来特别是后现代主义者的著作,那么就会丧失来自政府、公司和学术基金会的资助。这意味着,科学实验最终不能进行下去,那么科学研究对公共健康和生活水平的贡献就会急剧下降,让世界处于一种无知的状态。他们认为,这代表了一种倒退,不应该发生——无论目前的科学状况有多么不完美。

---

1. 彼得·格弗里-斯密斯:《科学哲学导论:理论和现实》,伊利诺伊州芝加哥:芝加哥大学出版社,2003 年,第 98 页。
2. 伊安·哈金:《科学革命》,牛津:牛津大学出版社,1981 年,第 1—2 页。
3. 汉纳·安德森、彼得·巴克、陈向:《科学革命的认知结构》,剑桥:剑桥大学出版社,2006 年,第 238 页。
4. 简·格林斯基:《创造自然知识:建构主义和科学史》,剑桥:剑桥大学出版社,1998 年,第 14 页。
5. 瓦索·金迪和西奥多·阿拉贝齐:"导论",载瓦索·金迪和西奥多·阿拉贝齐编

《重读库恩的〈科学革命的结构〉》,纽约:劳德里奇出版社,2012年,第2页。

6. 约翰·L.海尔布龙:"'数学家'的叛变与道德",载保罗·霍里奇等编《世界的变化:托马斯·库恩与科学的本质》,坎布里奇:麻省理工学院出版社,1993年,第112页。

7. 詹姆斯·A.马科姆:《托马斯·库恩的革命:科学的历史哲学》,伦敦:连续出版公司,2005年,第134—136页。

8. 金迪和阿拉贝齐:"导论",第2页。

9. 金迪和阿拉贝齐:"导论",第3页。

10. 彼得·诺威克:《崇高梦想:客观问题和美国历史行业》,剑桥:剑桥大学出版社,1988年,第536—537页。

11. 齐亚乌丁·萨达尔:"托马斯·库恩和科学战争",载理查德·阿皮尼亚内西编《后现代主义与大科学》,剑桥:艾肯图书公司,2002年,第216—221页。

12. 萨达尔:"托马斯·库恩和科学战争",第221页。

13. 巴里·巴尼斯:《T.S.库恩和社会科学》,伦敦:麦克米伦出版公司,1982年。

14. 大卫·布鲁尔:《知识与社会意象》第2版,伊利诺伊州芝加哥:芝加哥大学出版社,1991年,第7、166页。

15. 鲁诺·拉图尔和斯蒂芬·沃格:《实验室的生活:科学事实的建构》,纽约:塞奇出版公司,1979年,第27页。

16. 萨达尔:"托马斯·库恩和科学战争",第6页。

17. 艾伦·索卡尔:"一位物理学家的文化研究实险",《通用语》,1996年,第62—64页。

18. 引自迈克尔·博鲁比:"科学战争再现江湖",《民主杂志》第19卷,2011年,第67页。

11 当代印迹

要点

- 《科学革命的结构》提出了对科学知识的一种有争议的理解,它将继续在今天的社会科学和哲学领域引发争论和歧见。
- 学者们把库恩的著作放在逻辑实证主义*学派(强调形式的方法)和相对主义*学派之间(其假设是,解决科学问题最终的、"完美的"方法是不可能的),但它给每个人带来了挑战。
- 对库恩著作的回应延续到了21世纪,随着后现代主义*的兴起,学者们越来越多地质疑科学对其他学科有用的观点。

地位

1993年,德裔美国科学哲学家卡尔·亨佩尔*把托马斯·库恩的《科学革命的结构》誉为科学史*当中的里程碑。亨佩尔在一本科学哲学*文集开篇中,对库恩直截了当地说:"无论你的同事采取什么立场,汤姆,我相信他们都非常感激你的富有挑战性和启发性的思想。[1]"亨佩尔的赞颂特别值得关注,因为库恩的著作挑战了亨佩尔所采用的逻辑经验主义*方法。

许多学者发现,这部著作在科学史上的影响最小,而库恩恰恰希望它在此方面影响最大。事实上,最近的批评家声称"库恩在科学研究上的努力已经使科学史的重要性黯然失色,使政治讨论不能正常进行。[2]"

当然,这本以历史和文化为重点的著作,导致了逻辑经验主义的消亡。逻辑经验主义是一个建立在客观性和纯粹可验证分析观念基础上的学派,不过它对科学哲学的积极影响尚不明晰。虽然库恩著作这种前所未有的反响使其名声大噪,但也妨碍了《科学革命的结构》对库恩最重视的科学,即科学史,做出持续的、明确的贡献。在某种程度上,这是因为这种反响过多地关注库恩论点中的哲学内容——诸如范式*、不可通约性*和范例*等等——而不是库恩建立这些概念的基础:科学史文献。

> 50年前的这个月,20世纪最有影响的著作之一由芝加哥大学出版社出版了。很多普通人,即便不是大多数,可能从来没有听说过这本书《科学革命的结构》,也没听过它的作者托马斯·库恩,但是人们的思维几乎都受到了他思想的影响。关键在于,你是否曾经听说过或者使用过"范式转移",这个术语在当代组织变革和智力进步的讨论中使用得最多——甚至滥用。例如,用谷歌搜索它,会看到有超过10 000 000次的点击量。目前,它在亚马逊销售的图书中不少于18 300册。它始终是引用频率最高的学术著作之一。所以假如有一个伟大的想法像病毒一样传播,那就是它。
>
> ——约翰·诺顿:《托马斯·库恩:改变世人看待科学方式的人》

互动

《科学革命的结构》仍然是当代科学学术辩论的部分内容。但是,库恩对今天这种辩论的方向并不满意,这场辩论在科学现实主义者*(大致认为,我们通过观察获取科学事实)和建构主义者*(认为科学是一项解释工作,其中科学家的信念举足轻重)之间划出了战线。

两派之间的辩论发生在20世纪90年代,即所谓的"科学战争*"。在这个时期，现实主义的支持者——主要是科学实践者和科学哲学家——被抨击为"非理性"文化和后现代主义思想，经常受到政治左派人士控制。美国科学家保罗·格罗斯*和诺曼·莱维特*于1994年在这场"战争"中开了第一枪，当时他们出版了《高级迷信》。在这本书里，他们强烈批评后现代主义是"中世纪"的思想观念，谴责学术左派对科学的偏见[3]。在此之后，他们于1995年在纽约市举行了一次学术会"自科学与理性的飞翔"。大会上所宣读的论文对把科学批评当成"无稽之谈"、把批评家当成"江湖骗子"作了谴责[4]。

21世纪初，格罗斯、莱维特，以及那些攻击过学术左派的人，采取了一些措施结束这场争论。但是争论仍未平息，因为它与大学如何分配科学研究经费有关——金钱是这场"战争"的根源，就像它在现实生活中的许多冲突那样。对格罗斯和莱维特来说，挑战科学的具体结论损害了它的信誉，并使其在大学里边缘化了[5]。在学术左派里，美国哲学家迈克尔·博鲁比*认为，后现代主义者真正所反对的是科学"基要主义"不可碰论。通过划定反对这一观点的战线，各派都可以团结起来，因为格罗斯、莱维特和博鲁比都渴望有高质量的科学实践，并可以公开辩论。所有这些人都与库恩都有一个相同的愿望，即通过更好地理解科学的运作方式来提高科学在社会中的作用；所有的人与库恩不同的是，通过公开辩论，他们剥夺了其不可动摇的地位（库恩所守护的东西）。

尽管如此，科学在社会中的地位问题未必转化为明确的政治立场。博鲁比指出："20世纪对美国高等教育机构并不友好，学术抨击现在是美国增长最快的主要行业之一。[6]"21世纪，大学经费问题普遍受到诟病。本来学者们就对科学研究的本质争论不休，经费问题更给学校后勤工作增加了压力。

持续争议

《科学革命的结构》现在出到了第四版。最新的版本即50周年纪念版，再次引起人们对库恩著作的兴趣[7]。《结构》继续激发并促使人们思考，在通常情况下知识状态的根本变化。例如，即使一些学者的专业领域与库恩的科学史相差甚远——历史、人类学和社会学，他们也会从"革命结构"和"范式转换"的角度提出看法，如同库恩定义这个术语一样[8]。

在学术界之外，库恩本人也一直是出了名的、激进的反对者，他反对科学和学术行为中根深蒂固的规范。事实上，人们有时把他看作原始后现代主义者——早已在包括科学领域的弱势群体和边缘化群体中开始了后现代的研究兴趣。由于《科学革命的结构》在学术界以外的影响扩大，所以该书中的几个术语已经成了普通用语。这些术语在科学知识领域里引起了变革，其中包括科学革命的概念，但最有名的还要算"范式"和"范式转换"。现在我们同样能在大众文化和学术领域里听到这些术语。"范式转换"这个表达法在20世纪末确实成为网络流行语了，连网购倡导者都使用它来说服顾客购物习惯[9]。2001年，《大白痴的智能词汇指南》认为"范式转换"一词过度使用，以致变得毫无意义[10]。

---

1. 卡尔·亨佩尔："托马斯·库恩：同事和朋友"，载保罗·霍里奇等编《世界在

变化：托马斯·库恩和科学的本质》，坎布里奇：麻省理工学院出版社，1993年，第7—8页。
2. 艾斯特-米丽娅姆·森特："评史蒂夫·福勒、托马斯·库恩：当代哲学史评论"，《政治评论》第63卷，2001年第2期，第392页。
3. 保罗·格罗斯和诺曼·莱维特：《高级迷信：学术左派及其对科学的思想观念》，巴尔的摩：约翰斯·霍普金斯大学出版社，1994年。
4. 齐亚乌丁·萨达尔："托马斯·库恩和科学战争"，载理查德·阿皮尼亚内西编《后现代主义与大科学》，剑桥：艾肯图书公司，2002年，第189页。
5. 彼得·格弗里-斯密斯：《科学哲学导论：理论和现实》，伊利诺伊州芝加哥：芝加哥大学出版社，2003年，第146页。
6. 迈克尔·博鲁比和卡里·纳尔逊编《受抨击的高等教育：政治、经济和人文危机》，纽约：劳德里奇出版社，1995年，第1页。
7. 托马斯·尼克尔斯："导论"，载托马斯·尼克尔斯编《托马斯·库恩》，剑桥：剑桥大学出版社，2003年，第1—19页。
8. 卡罗琳·麦茜特："生态革命的理论结构"，载查尔·米勒和哈尔·罗思曼编《脱离困境》，匹兹堡：匹兹堡大学出版社，2014年，第18—27页。
9. 肯特·吉尔曼："十大流行语"，访问美国CNET科技资讯网2013年8月17日内容，网址：http://www.cnet.com/1990-11136_1-6275610-1.html。
10. 保罗·麦克费德里斯：《大白痴的智能词汇指南》，纽约：阿尔法，2001年，第142—143页。

12 未来展望

要点

- 科学是不偏不倚的知识积累,而有人攻击科学地位的可信性,如何应对这一论争?《科学革命的结构》一书将继续发挥作用。
- 学界人士可能会继续争论库恩著作的意义,及其对科学史和科学哲学*的重要性。
- 《科学革命的结构》对于科学知识和实践提出了激进而富于创意的理解。本书改变了科学家的公众形象,将"范式*"这一概念变为一个寻常词汇。本书也激发了一代代科学家和学者,使他们反思科学知识的地位。

潜力

自 1962 年首次出版以来的半个世纪,托马斯·库恩的《科学革命的结构》彰显了科学史*的转变。据库恩本人观察,人们赞扬他的著作,不仅仅是因为它在科学史革命中发挥的作用。这种赞誉方兴未艾,而本书的许多方面也有待重新解读[1]。

库恩著作的潜力在于,它对当代社会学*两个流派的论争非常有用:科学知识*社会学(认为社会条件创造科学知识)和建构主义*社会学(认为科学家不是直接研究现实,而是由实验结果"建构"现实)——本书对后现代主义*与科学多元文化*评论也很有用。另外,考虑到科学家的工作环境,本书继续质疑科学平衡内在论*读物的可能性。

库恩并没有打算刻意推动这些发展。事实上,他只想为少数科学家、历史学家、科学哲学家读者写一本书。但只要学者们继续辩论，库恩对积累的、线性的、科学发展的批评就仍然保持其影响力。

我们可以把库恩的著作看作新观点的源泉，这些观点辐射出去，产生了他本人也未能预见的效果。可以合理预测，它们还将继续传播发展。我们也不禁好奇，在何种程度上它们还可以算作库恩的精神遗产。这也将决定，这些领域的学者和实践者是否需要回到原文，搜寻新观点并加以阐发²。

> "科学战争"议题旨在回应对女权主义者、多元文化主义者及科学社会批评者的"强烈反对的尖刻语气"……强烈抵制并吓退任何这样的人，他们敢于质疑科学的性别负载假设、科学经验主义的资本基础，以及科学技术对社会和环境的破坏性影响。
>
> ——齐亚丁·萨达尔：《托马斯·库恩和科学战争》

未来方向

在库恩派思想缺席的情况下，最有可能发掘《科学革命的结构》潜力的人是同一批学者。在2012年《科学革命的结构》50年周年纪念版准备出版阶段，他们都对库恩做出了积极回应。

其中有美国历史学家、社会学家史蒂夫·福勒*，他写了《托马斯·库恩：当代哲学史》(2000)，书中的核心问题是："库恩的观点有多激进？"他的答案很响亮："不太激进。"英国科学哲学家亚历山大·伯德*所著的《托马斯·库恩》(2000)是通过《结构》一书的核心概念折射来研究库恩的；既从个体角度考虑，如"范式"和"不可通约性*"，也从整体的科学哲学研究背景下考察。借此契机，美国科学哲学家托马斯·尼克尔斯*搜集了库恩生平、时代和著作的资料，编写了名为《托马斯·库恩》（2003）的文集。意大利科学史学家斯特凡诺·加泰伊*在其著作《托马斯·库恩的语言学转向及逻辑经验主义遗产》（2008）中，着力探究库恩著作与逻辑经验主义*（据称这是库恩著作攻击的对象）的关系。

《重读库恩的科学革命的结构》（2012）是由希腊科学历史学家瓦索·金迪*和西奥多·阿拉贝齐*编辑的，盘点了受库恩激发所产生的最新研究动向。他们收集的文章特别讨论科学研究中概念的功能、逻辑实证主义*、历史与科学哲学的关系，以及科学进步的本质。

尽管研究库恩的著作之多令人惊讶，它们的重点都放在其观点上，而没有试图运用。所以库恩的著作很可能只能成为思想史*的一章。伊安·哈金*在《科学革命的结构》纪念版中解释道："正因为《结构》是一部伟大的著作，人们可以以无数种方式解读，以无数种方式应用。[3]"库恩的著作所启发的未来思想，将如同《结构》1962年首次面世就给学术界带来惊喜一样，激发很多有价值的观点。

小结

《科学革命的结构》值得一读，有三个相互关联的理由：

第一，这本著作改变了我们对科学本质和特点的理解。它使用了一些已进入日常生活语言的概念做到了这一点，也许最为人们熟知的就是范式和范式转换*。《结构》中讨论的这些概念及其他一些概念，将继续在世界各地的家庭、学校和研讨室中被人们使用。

第二，库恩的著作激发了科学史以外其他领域的巨变，影响了哲学和一系列社会科学领域的思想观念。他的科学革命观念已在很大程度上等同于人类的奋斗与发现的方式，这种方式提高了人类在世界上生存与繁荣的能力。

第三，库恩在其著作中所采用的严苛概念工具在批判性思维方面提供了很好的教材，帮助各学科的读者培养批判能力。这是本书对科学之外领域产生的另一个方法：即使商业或知识产业的专业人士也可以从库恩的思想中受益。从这个意义来讲，库恩的著作可能是20世纪用途最广泛的书籍之一。即便我们步入在21世纪，其作用也将如此。

---

1. 托马斯·S.库恩：《结构以来的路：1970—1993哲学论文集与自传式访谈》，詹姆斯·科南特和约翰·海于格兰编，伊利诺伊州芝加哥：芝加哥大学出版社，2000年，第90—91页。
2. 玛尔妮·休斯-沃灵顿："托马斯·塞缪尔·库恩"，载《历史上五十位最重要的思想家》，伦敦：劳德里奇出版社，2003年，第191—192页。
3. 伊恩·哈金：《导读》，载于托马斯·库恩著《科学革命的结构》第4版，伊利诺伊州芝加哥：芝加哥大学出版社，2012年，第VIII页。

术语表

1. **分析命题**：因其本身意义而真实的命题，通常是广为接受的事实陈述，如："乌鸦是黑的"。

2. **分析-综合区分**：哲学家用来区分事实或本真命题（分析命题），和由于对世界的陈述而被认定真实的命题（综合命题）的方法。

3. **异常现象**：在库恩的范式概念背景下，指范式提供的解释与范式拟解释的现实各方面的差别。它使人们对导致危机科学和科学革命阶段的范式产生疑问。

4. **古典力学**：应用力学，涉及物体和力的运动和平衡，其分支包括运动学、动力学和静力学。

5. **认知科学**：分析人们如何通过视觉、听觉、味觉和触觉等，获取和使用接到的信息。

6. **认知世界观**：人们将抽象概念，通过与视觉、听觉、嗅觉、味觉和触觉积累的对现实的感知联系起来看待世界的方式。例如，从视觉和触觉，我们感知自己和多数物体一样站在地上；而重力的概念使我们知道能够站立的原因，这是仅靠感官无法得知的。

7. **冷战**：美苏军事紧张时期，自二战结束至1991年苏联解体为止。

8. **共识范式**：危机科学时期，科学家从竞争性选项中选择的范式，一种引发科学革命的选择。

9. **建构主义者**：在科学哲学中，被称为建构主义学派的信奉者。

10. **建构主义**：认为科学家不是直接研究现实，而是通过他们在实验中搜集的信息建构对于现实的理解。

11. **危机科学**：在一种不确定阶段/时期，始于现实与一种范式间的反常现象，对共识范式提出疑问。如果可以通过解释消除反常现象，而只需对共识范式作很小的修改，这一阶段可能导致科学革命（届时选择一种新的共识范式），或对常规科学的修正。

12. **达尔文理论**：物种进化理论，认为动物和有机体继承父母那些有利于竞争、生存繁衍的心理和生理特点，从而得到进化。

13. **学科基质**：在库恩著作中，科学界在其专业领域内使用的符号概括、模型、数值和解决办法。

14. **实证的**：对可观察证据而非对假设或理论进行研究。

15. **实证主义**：认为一切知识都来自于我们通过感官（视觉、听觉、嗅觉、味觉和触觉）获得的对现实的经验。因此，在科学里，实证主义强调为了"看到"现实，必须进行准确的实验。

16. **实证主义者**：实证主义的信奉者，在科学领域内倡导为"看到"现实而进行准确的实验。

17. **深奥**：用于形容对任何没有预备知识的读者和观众来说难以理解的语言、思想流派或理解体系，只适用于具有专门知识的内部人士。

18. **范例**：常规科学时期，共识范式下对科学问题的著名的解决办法，成为所有科学家模仿典范所遵循的范式。

19. **外在主义**：一种将科学理解为科学界和科学实验以外因素（如社会因素、政治事件、经济因素）之产物的方法。

20. **非常规科学**：科学危机与科学革命之间的阶段。多种范式争相解释反常现象，其太严重以至现存共识范式无法继续存在；因科学活动陷入混乱局面，故称为非常规科学。

21. **女权主义**：一种政治与学术议题。提倡女性具有自愿选择任何生活角色的平等权利和自由，即使这些角色传统上专属男性。

22. **格式塔心理学**：认为对世界的感知是"格式塔"的心理学派。认为当头脑用理论框架（如概念）将对世界的感知（景象、声音、语言）联系起来，就产生了思想和印象，即格式塔。格式塔心理认为我们通过整理思想，让独特的、新的经验突然改变我们对什么是"真"的理解，从而建立世界观。

23. **格式塔转换**：根据格式塔心理学，它指个人的感知由一个格式塔变为另一个的时刻。格式塔转换之后，人们看待世界的方式和以前不再相同。

24. **全球化**：过去30年各大公司、经济体及国家实行对外开放，进行国际交流，创建新的大众市场的进程。

25. **哈佛大学**：美国顶尖私立研究型大学，创建于1636年，位于马萨诸塞州坎布里奇市。

26. **广岛**：位于日本本州岛西部的城市。1945年二战接近尾声时，美国空军向日本两个城市投掷原子弹，广岛就是其中之一。

27. **思想史**：历史学分支，研究人们自古以来如何形成和改变对世界的看法，并将这些看法作为社会、文化、经济、政治生活的一种功能。

28. **科学史**：历史学分支，主要阐释自古以来科学的发展。

29. **科学史学会**：该学会提升全美对科学史及其社会文化关系的兴趣。主席由行政办公室选举产生，其候选人由同行按学术成就挑选。

30. **理想主义**：哲学思想流派，认为现实是人类思想的构建，而非可以直接调查的外部事物。

31. **不可通约性**：范式的一个特点，不管以何种方式，各范式之间不能完全理解或借用对方的语言、概念或方法。每个领域都有自己的范式，因此不同领域的科学家无法轻易跨越范式进行合作，当新的共识范式出现，同领域科学家就无法再使用被其推翻的旧范式。

32. **内在主义**：科学哲学史上一种立场，单纯从科学家行为与科学实践本质阐释科学。内在主义者并不考虑外部环境。

33. **非理性**：就任何课题思考和得出结论时，没有根据广为接受的逻辑标准形成合理的解释。理性主义的反面。

34. **逻辑经验主义**：20世纪初期盛行的科学哲学运动。逻辑实证主义认为，科学应分析现实的各个方面，以理解整体。还认为一切科学家的结论，可以通过让其他科学家重复实验过程而得以证实。

35. **逻辑实证主义**：20世纪30年代初维也纳学派哲学家的理论与学说，认为形而上学的、非理性的、推测性问题的逻辑基础都很薄弱。逻辑实证主义旨在逐步形成形式化方法——类似于数学科学的方法，以哲学语言来证实经验性质的问题。

36. **曼哈顿计划**：1942—1946年间，由美国主持设计和产生的首个核武器研究计划，得到加拿大与英国支持。

37. **麻省理工学院**：位于马萨诸塞州坎布里奇市，创建于1861年的私立研究型大学，最初以工程学和有关美国经济科技进步的课题为研究重点。

38. **多元文化主义**：为鼓励同一社会中不同文化并存而设计的原则。

39. **长崎**：日本九州岛城市。1945年二战接近尾声时，美国空军向日本两个城市投掷原子弹，长崎就是其中之一。

40. **常规科学**：共识范式统辖科学家研究领域的时期。随着实验进展，科学家也有可能改进这一范式。

41. **规范（的）**：根据理想标准或基准，而非实际情况进行判断或测量的文字陈述。

42. **1973年石油危机**：当时埃及和叙利亚领导的阿拉伯国家对以色列发动了中东战争，美国介入战争后，这些国家对美国发起禁运，使油价由3美元上升到12美元。此危机导致西方国家生活费用急剧上涨。

43. **苍白**：任何缺乏深度、色彩浓度的物体或有机体。

| 术语表

44. **范式**：在库恩对科学的理解中，指科学家达成共识的一种概念，以此来解决科学问题，引领未来科学研究，直到发生范式转换。

45. **范式转换**：当一种范式达到不可通约的状态，科学家对新的范式达成共识，这表明他们从旧范式转而忠实于新范式。

46. **科学哲学**：哲学分支，着力解释科学家进行工作的基本原则。

47. **后殖民研究**：人文学科分支，意在修正人类学、历史、社会学（及其他学科），从而将曾遭受殖民统治的个人与群体的视角纳入其研究。往往包含受欧洲列强殖民统治的非洲、亚洲、南美及其他地区。

48. **后现代主义**：人文学科和社会学的运动。20世纪80年代末生根发芽，促使学者探讨以前未被代表的群体在创造历史中的作用。这些群体包括妇女、少数民族、少数宗教以及边缘化性取向者。

49. **前共识科学**：科学活动的最早阶段，敌对学派的科学家争相解释各种科学现象，直到选择了第一个大家都在研究中遵循的范式，常规科学的第一个时期随之到来。

50. **普林斯顿大学**：创立于1746年，位于新泽西州，美国领先的私立研究型大学。

51. **量子机械学**：物理学分支，研究最小颗粒的运动和相互作用。在物理世界中专注于更小规模，加深了古典机械学的发现。

52. **量子物理**：物理学分支，研究物理世界中最小层级的物体：原子和原子微粒。

53. **理性主义**：哲学思想流派，注重使用逻辑和理性思维去解决问题，而非依靠直觉或信仰。

54. **现实主义**：科学哲学思想流派，认为科学家可以直接接触和描述现实，而不以他们的观点或惯例影响结论的真实性。

55. **相对主义**：科学思想流派，认为科学家做研究时只是得出暂定结论，而非永久不变的绝对最终结果。

56. **革命科学**：当现实中出现原有范式不能解释的反常现象，科学家开始对其存疑，而几种新的范式竞相成为科学界新的共识范式的时期。

57. **科学战争**：20 世纪 90 年代及 21 世纪初的一系列思想交流活动，当时科学现实主义和后现代主义批评家就科学的性质展开了讨论。

58. **科学界（科学共同体）**：任一时期世界上所有科学家的总称。

59. **科学革命**：不时发生的、导致科学知识发展的过程，往往伴随着新发现。革命期间，科学家摒弃旧范式，遵照新范式，因它能更好地解释现实。

60. **社会学**：对人类社会结构与历史的研究。

61. **科学社会学**：社会学分支。探讨科学家的社会、经济、职业地位是如何帮助形成科学圈中的研究课题和结论的。

62. **科学知识社会学**：科学社会学一个思想流派，从库恩的研究中获得启发，主张对科学作建构主义阐释。

63. **苏联**：苏维埃社会主义共和国联盟，存在于 1922—1991 年期间。权力中心位于俄罗斯，但也包含其他许多加盟国，如乌克兰和格鲁吉亚。

64. **综合命题**：一个命题因它与世界关联而为真，而非因其是事实陈述。如：乌鸦袭击小鸟。此命题对乌鸦并不普遍适用，但可以通过对大多数乌鸦的观察得到证实。

65. **隐性知识（默会知识或内隐知识）**：通过简单的说或写等行为不能轻易传播给另一人的知识，包括往往被视为理所当然的一些假设。

66. **目的论**：对一个科目的学习，如科学史，最终达到某一目的或终极目的。目的论往往招致批评，因其过于注重目的，而忽视通往科学现有形式的混乱过程。

67. **理论物理**：物理学分支，依靠理论、数学及抽象推理等对自然世界提出可能的解释。因研究主体的大小和性质，其结论无法在实验室验证。例如，人们无法在实验室检验宇宙的性质。

68. **观念单元**：对世界的基本观念和感知。如亚瑟·O.洛夫乔伊所解释，如果不与思考结合而形成世界观，其本身并无意义。比如，"天空是蓝色"这一概念包含的观念单元有：能够看见天空，能够将蓝色与其他颜色区分，认识"天空"和"蓝色"这两个词等。

69. **可证明意义理论**：主张为证明命题或科学结论的意义与真实性，科学家应具备再创造一个由同行完成的实验、依然得出相同结论的能力。如果同一实验得出不同结论，即证明其同行实验是错误的。

70. **实证主义原理**：科学实验为保证真实性，其结论必须可由第三方复现的规则，即第三方重复相同实验仍可得出相同结论。

71. **加州大学伯克利分校**：加州大学体系中顶尖的大型公立研究型大学，其教学和研究均在国际上享有盛誉。

72. **第一次世界大战（1914—1918）**：协约国（英国、法国、俄罗斯和美国等）与同盟国（德意志、土耳其和奥匈帝国等）之间的全球性武装冲突。

人名表

1. **汉纳·安德森**：丹麦哲学家，任职于奥尔胡斯大学科学技术学院数学与科学系。她深入研究了库恩的著作，致力于把科学哲学与科学教育相结合的工作。

2. **西奥多·阿拉贝齐**：希腊雅典大学哲学和科学史副教授。他最近出版了《代表电子》一书（2006）。

3. **弗朗西斯·培根**（1561—1626）：英国哲学家、政治家、律师和作家。他的主要著作是《新工具》（1620）和《新大西岛》（1627）。

4. **彼得·巴克**：英国历史学家和科学哲学家。他目前供职于美国俄克拉荷马大学，任科学史教授。

5. **巴里·巴尼斯**（1943年生）：英国科学哲学家，目前是埃克塞特大学社会学教授。自20世纪70年代至90年代，他在爱丁堡大学与大卫·布鲁尔从事科学知识的社会学重大项目研究。

6. **乔治·贝克莱**（1685—1753）：英国哲学家，是英国逻辑经验主义哲学主要倡导者之一。这个哲学分支认为知识来自经验，而不是理性。他的主要著作是《人类知识原理》（1710）。

7. **迈克尔·博鲁比**（1961年生）：美国哲学家，美国宾夕法尼亚州立大学人文学院院长、教授，著有《受抨击的高等教育》（1995）。

8. **亚历山大·伯德**：英国科学哲学家，布里斯托尔大学哲学教授，著有《托马斯·库恩》（2000）和《科学哲学》（1998）。

9. **大卫·布鲁尔**（1942年生）：英国社会学家，爱丁堡大学科学技术教授。他创立了科学知识的社会学重大研究项目，著有《知识与社会意象》，并因此名声大噪。

10. **J. Z. 布赫瓦尔德**（1949年生）：美国科学史学家，目前是加利福尼亚州帕萨迪纳市加州理工学院的多丽丝和亨利·德雷弗斯历史学教授，著有大量科学史论著，包括一部物理学实践专辑《科学实践：研究物理的理论和故事》。

11. 鲁道夫·卡尔纳普（1891—1970）：德国逻辑经验主义科学哲学家。他 1935 年移民美国，主要著作是《哲学与逻辑句法》（1935）。

12. 陈向：亚裔美籍科学哲学家，目前是加州路德大学哲学教授。此学科主要贡献是安德森、巴克和陈向合著的《科学革命的认知结构》（2006）。

13. 詹姆斯·布莱恩特·科南特（1893—1978）：美国化学家，哈佛大学校长，美国驻西德首位大使。

14. 尼古拉·哥白尼（1473—1543）：文艺复兴时期波兰数学家和天文学家，他创立了现行的宇宙日心说，地球绕着太阳转。

15. 阿瑟·丹托（1924—2013）：美国艺术批评家和哲学家，在许多领域里都有成就，特别是史学理论。最为人们所熟知的是，他是美国最古老周刊《民族报》的艺术评论员。

16. 查尔斯·达尔文（1809—1882）：英国自然科学家，提出生命同宗、物种演化源自自然选择、适者生存等理论。

17. 皮埃尔·迪昂（1861—1916）：法国历史学家和科学哲学家，其代表作是《世界体系论》（1913—1916），书中著名的论述为中世纪与早期现代科学的延续。

18. 保罗·费耶阿本德（1924—1994）：出生奥地利，移居美国，是反传统科学哲学家，主要著作为《反对方法》。

19. 米歇尔·福柯（1926—1984）：法国哲学家、历史学家、社会理论家、语文学家、心理学家、20 世纪最著名的思想家之一。他的主要著作是《知识考古学》（1969）。

20. 迈克尔·弗里德曼（1947 年生）：美国科学哲学家，执教于斯坦福大学，并从事研究工作。其著作《理性动力》发展并探索了库恩范式转移概念等领域，而库恩本人对此却论述不足。

21. 史蒂夫·福勒（1959 年生）：美国哲学家和社会学家，主要著作是《库恩与波普尔》（2003）和《科学与宗教》（2007）。

22. 斯特凡诺·加泰伊：意大利科学哲学家，意大利卢卡高级研究所研究员。其博士论文是研究库恩著作的文章：《托马斯·库恩的不完整革命》（2007）。

23. 简·格林斯基（1957年生）：美国科学哲学家，新罕布什尔大学历史人文学教授，《创造自然知识》的作者。

24. 保罗·格罗斯：美国生物学家，与诺曼·莱维特合著有《高级迷信》（1994），并著有《神创论的特洛伊木马：智能设计的楔子》（2004）。

25. 伊安·哈金（1936年生）：加拿大著名历史学家和科学哲学家，主要著作是《概率的突现》（1975）。

26. 诺伍·德罗素·汉森（1924—1967）：美国科学哲学家，其著名论断是：当科学家观察世界时，他们对所看到的强加了许多预设理论。

27. 约翰·L.海尔布龙（1934年生）：美国科学史学家，加利福尼亚大学伯克利分校历史学教授，主要著作是《十七、十八世纪电学：早期现代物理学研究》（1979）。

28. 卡尔·亨佩尔（1905—1997）：德裔美国科学哲学家，是著名的逻辑经验主义者，主要著作是《科学解释》（1967）。

29. 理查德·霍夫斯塔特（1916—1970）：美国历史学家，研究方向为美国现当代历史，其著作广泛涉及思想史和历史。

30. 保罗·霍伊宁根–休内（1946年生）：德国科学哲学家，因其对库恩进行新康德主义解释而著名，主要著作是《重构科学革命：托马斯·S.库恩的科学哲学》（1993）。

31. 大卫·休姆（1711—1776）：英国哲学家，英国经验主义哲学的主要倡导者之一；这一哲学流派认为，知识源自经验，而不是理性；其主要著作是《人性论》（1739）。

32. 伊曼努尔·康德（1724—1776）：德国哲学家，现代最重要的思想家之一，主要著作是《纯粹理性批判》（1781）。

33. 约翰尼斯·开普勒（1571—1630）：德国数学家和天文学家，在改变17世纪科学家对世界的认识起了重要的作用；他的主要理论贡献是为我们提供了认识行星运行规则的模型。

34. 瓦索·金迪：希腊科学哲学家，希腊雅典大学哲学和科学史助理教授。到目前为止，他的主要著作是《库恩与维特根斯坦：科学革命结构的哲学考察》（1995）。

35. 亚历山大·柯瓦雷（1892—1964）：在美国的法裔俄国流亡者，对17世纪科学史进行了杰出的分析，帮助创造了"科学革命"一词，用来描述学术变革时期。其主要著作是《从封闭世界到无穷宇宙》(1957)。

36. 伊姆雷·拉卡托斯（1922—1974）：匈牙利科学哲学家和数学家，主要著作是《科研纲领的方法论》(1978)。

37. 布鲁诺·拉图尔（1947年生）：法国科学社会学家和人类学家，主要著作是《行动中的科学》(1987)和《实验室的生活》(1979)，以讨论因社会环境的影响减少科学家工作而著称。

38. 安托万·拉瓦锡（1743—1794）：法国化学家，被誉为"现代化学之父"，18世纪末"化学革命"的领导者。

39. 诺曼·莱维特（1943—2009）：美国数学家，执教于美国罗格斯大学，因其著作《高级迷信》(1994，与保罗·格罗斯合著)为人所知。

40. 约翰·洛克（1632—1704）：英国经验主义哲学家，被誉为"古典自由主义之父"，主要著作是《人类理解论》(1690)。

41. 亚瑟·O.拉夫乔伊（1873—1962）：德裔美国流亡者，因其著作《存在巨链》(1936)闻名，该书开启了思想史的现代研究。

42. 恩斯特·马赫（1838—1916）：德国哲学家，《力学科学》(1883)的作者，该书主要论述了关于形而上学的经验主义。

43. 安内利泽·迈尔（1905—1971）：德国科学史学家，她的论著选集是用英语出版的，即《论精确科学的起点：安内利泽·迈尔关于中世纪晚期自然哲学的文选》(1982)。

44. 詹姆斯·A.马科姆：美国科学哲学家，执教于德克萨斯州韦科市贝勒大学。他研究了科学发展，近期撰写了《系统生物学的概念基础：导论》(2009)。

45. 玛格丽特·玛斯特曼（1910—1986）：英国语言学家和哲学家，因对计算语言学和计算机自动翻译的开拓性研究而闻名。

46. 海琳·梅斯热（1889—1944）：法国著名的历史学家和科学哲学家，她的主要著作是《十七世纪初至十八世纪末法国化学学说》(1923)。

47. 艾兰·马斯格雷夫（1940年生）：新西兰科学哲学家，卡尔·波普尔的学生，《常识、科学和怀疑主义》（1992）的作者。

48. 奥图·纽拉特（1882—1945）：奥地利逻辑实证主义哲学家，维也纳科学哲学界领袖。纽拉特的主要英文版著作是《经验主义和社会学》（1973）。

49. 艾萨克·牛顿（1642—1727）：英国物理学家和数学家，是运动定律和现代引力理论的开拓者，因其著作《数学原理》（1687）闻名。

50. 托马斯·尼克尔斯：美国科学哲学家，里诺市内华达大学基金会教授；撰写了很多论著研究库恩和科学史的其他领域。

51. 乔治·奥威尔（1903—1950）：埃里克·亚瑟·布莱尔的笔名，英国小说家和记者；因其反乌托邦小说《一九八四》和中篇小说《动物庄园》闻名，两部小说谴责了专横的政府。

52. 让·皮亚杰（1896—1980）：瑞士心理学家和哲学家，提出了著名的儿童认知发展理论，主要著作有《儿童智慧的起源》（1953）和《儿童的现实构建》（1955）。

53. 马克斯·普朗克（1858—1947）：德国理论物理学家，被认为是量子理论的奠基人，因此理论获得1918年的诺贝尔物理奖。

54. 迈克尔·波兰尼（1891—1976）：英国裔匈牙利科学哲学家和化学家，著名著作有《个人知识》（1958）和《隐性维度》（1966）。

55. 卡尔·波普尔（1902—1994）：英裔奥地利科学哲学家，被广泛誉为20世纪伟人之一，他的名作是《科学发现的逻辑》（1934）。

56. 威拉德·冯·奥曼·蒯因（1908—2000）：美国极有影响的哲学家和逻辑学家，主要著作是《词语和对象》（1960）。

57. 杰瑞·莱文兹（1929年生）：美国科学哲学家、环境顾问，致力科研的不确定性和道德伦理研究，并因此闻名。其著作《科学无意义指南》（2005）对这个领域有杰出贡献。

58. 乔治·A.赖施：美国科学史学家，因其对科学的外部性研究闻名，如《冷战是如何改变科学哲学的》（2005）。

59. 霍华德·桑克：澳大利亚科学哲学家，现为墨尔本大学哲学助理教授，著有《不可通约论》（1993）。

60. 齐亚丁·萨达尔（1951 年生）：英国著名作家、学者、公众人物，著有许多著作，包括《读古兰经》(2011) 和《托马斯·库恩和科学战争》(2000)。

61. 乔治·萨顿（1884—1956）：比利时出生的美国化学家和科学史学家。他试图撰写九卷科学史，可是在去世时只完成了三卷。

62. 达德利·夏皮尔：美国科学哲学家、维克森林大学教授，因其严厉批评托马斯·库恩和保罗·费耶阿本德而为人所知。

63. 艾伦·索卡尔（1955 年生）：伦敦大学学院数学教授、纽约大学物理教授。

64. N. M. 斯沃德罗（1941 年生）：芝加哥大学荣誉退休的历史学、天文学和天体物理学教授，加州理工学院访问教授。

65. 哈利·S. 杜鲁门（1884—1972）：美国政治家，第 33 任总统，任期时间是第二次世界大战随即结束的 1945—1953 年，任职期间冷战开始。

66. E. O. 威尔森（1929 年生）：美国生物学家、自然科学家和作家，是世界上首屈一指的蚂蚁专家，也是哈佛大学退休教授，其近期著作是《人存在的意义》(2014)。

67. 诺顿·怀斯（1940 年生）：加利福尼亚大学洛杉矶分校著名历史学教授，近期主要研究自 18 世纪至今的科学和工业化。

68. 路德维希·维特根斯坦（1889—1961）：被誉为 20 世纪最伟大的哲学家，他的主要著作是《逻辑哲学》(1921) 和《哲学研究》(1953)。

69. 斯蒂芬·沃格（1950 年生）：英国社会学家，目前为牛津大学赛德商学院社会科技研究中心负责人，与鲁诺·拉图尔*的合著了《实验室的生活》(1979)。

70. 约翰·沃勒尔（1946 年生）：伦敦经济学院科学哲学教授，是伊姆雷·拉卡托斯的学生，著有《科学本体论》(1994)。

71. 约翰·M. 齐曼（1925—2005）：英国出生的物理学家，定居新西兰；其研究在凝聚态物理学领域处于行业领先地位；作为科学代言人，他在科学领域之外也具有最广泛的声誉，他的《真正的科学》(2000) 是这个话题家喻户晓的著作。

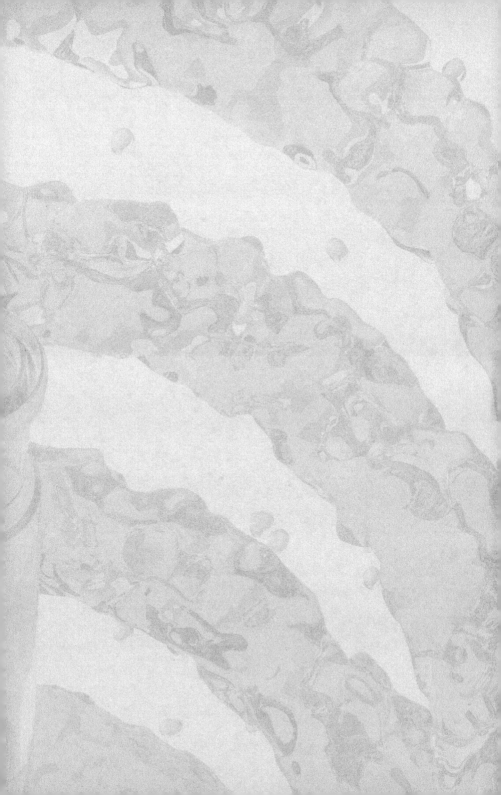

WAYS IN TO THE TEXT

KEY POINTS

- Thomas Samuel Kuhn (1922–96) was an American physicist, historian, and philosopher of science.
- In *The Structure of Scientific Revolution*s (1962) he argued that all advances in scientific knowledge resulted from "revolutions" in scientific knowledge.
- The book transformed the notion of science and the way in which science develops as a practice shaped by scientists and their methods.

Who Was Thomas Kuhn?

Thomas Kuhn was born in 1922 in Cincinnati, Ohio. His parents were Samuel L. Kuhn, an industrial engineer, and Minette Stroock Kuhn, whom he would later describe as "the intellectual in the family."[1] He published *The Structure of Scientific Revolutions* in 1962, when he was 40 years old.

By the time Kuhn graduated from high school, he knew that he wanted to pursue mathematics and physics. After attending boarding school on the east coast of the United States, he enrolled at Harvard University* in 1940. Once he had received his undergraduate degree, he spent two years in the military working on radar technology.

America's decision to deploy atomic bombs in 1945 near the end of World War II, destroying two Japanese cities, reminded the world that science's power to shape human destiny was an awesome responsibility. Kuhn decided to address this by going beyond the applications of science to study the principles underlying science itself.

After the war, Kuhn resumed academic study and in 1949 gained his PhD from Harvard, where the university's president, the scientist James Bryant Conant,* became his mentor. Conant, who had helped develop the atomic bomb, enlisted many renowned scholars in an effort to enhance the popularity of science. Thomas Kuhn, then unknown, was one of them.

Largely because of Conant's influence, Kuhn made teaching the history and philosophy of science* his areas of expertise. Before Kuhn published *The Structure of Scientific Revolutions*, the history of science* was seen as a sideline—interesting, but not worthy material for an academic discipline. Kuhn's work changed that. It also brought a number of scientific concepts into the mainstream, including "paradigm"* (roughly, a model of knowledge that explains the results of scientific experiments), "scientific revolution"* (the moment one paradigm gives way to another), and "paradigm shift"* (the intellectual and scientific consequences of a scientific revolution).

What Does *The Structure of Scientific Revolutions* Say?

Scientists had always seen their work as linear, with each new advance adding to and enhancing the store of scientific knowledge amassed since ancient times. In *The Structure of Scientific Revolutions*, however, Thomas Kuhn argued that science was cyclical. That is, progress occurs in repeating phases.

As Kuhn saw it, a period of "normal science"* would end when scientists overturned the prevailing paradigm, creating a period of "revolutionary science."* Rather than building on established stores of knowledge, the breakthroughs made in this revolutionary period permanently change the way scientists understand the world. In addition to the concepts of "paradigm" and "revolutionary science," Kuhn introduced a third concept: "incommensurability."*

Scientists overturning a prevailing paradigm are required to replace it. The new paradigm, offering a different and sometimes opposing account of reality, will be "incommensurable"—that is, the two paradigms have no common or shared means of explaining reality. So the selection of a "consensus paradigm"* at the end of a scientific revolution is as much a matter of choice by scientists as it is based on the power of the paradigm to explain reality.

If scientists in any given era see the world in the same way, Kuhn claims, it is because they use the same paradigm. But that does not mean the world is exactly as they observe it. The world's reality exists independently of the observer. When the paradigm changes, it is not the *world* that changes but the *scientist*, who now observes the world from a different perspective. This is a point some readers of Kuhn have failed to grasp, believing instead that he meant "paradigm shift" to refer to a world-altering event.

Kuhn turns to history to validate his theory. In the eighteenth century, the French chemist Antoine Lavoisier* discovered oxygen. Oxygen pre-existed this discovery, of course; humans had been breathing it for a very long time before Lavoisier's birth. Lavoisier's discovery, Kuhn explains, changed the paradigm. The world had not changed. But scientists never saw it in the same way again.

The Structure of Scientific Revolutions retains a wide readership more than 50 years after its initial publication. By 2003, the work had sold over a million copies and been translated into at least 20 different languages. *Structure* has indeed changed the way scientists look at their endeavors. Although some disagree with Kuhn's conclusions—indeed, one long-running disagreement was labeled "the science wars"*— *Structure* attracted readers inside and outside of science because it captured a spirit of questioning that seemed to pervade society in the early 1960s. There was a "paradigm shift" in the air, it seemed. So Kuhn's work about shifting ideas in science fitted perfectly with wider discussions taking place.

The academic discipline of sociology* (the study of the structure and history of human societies) grew more popular during this period. Sociologists had always looked at how social conditions shaped ideas; now they studied how social conditions had shaped science and historians of science.

The general public was also highly attuned to science during this period. Many, such as Kuhn's mentor James Bryant Conant, saw science as key to winning the tense standoff of the Cold War* between the United States and the Soviet Union.* But increasing numbers of people struggled to reconcile that goal with the potential devastation of a nuclear war. People questioned the role that scientists played in developing such destructive weapons.

Structure served as a springboard for these ideas. Some of the conclusions people drew from the work surprised Kuhn. A few even worried him. He had, after all, been writing about the internal structure of science, not about the power of ideas to change the world. But there was no doubt that Kuhn's work opened the door to questioning. And that door will likely not close again.

Why Does *The Structure of Scientific Revolutions* Matter?

Thomas Kuhn's idea of "revolutionary" periods of science upended centuries of established wisdom. Before Kuhn, scientists believed that they were detached observers, building upon the objective observations and discoveries of the scientists who had come before. After Kuhn published *The Structure of Scientific Revolutions*, those comforting certainties no longer applied.

In that sense, *Structure* is itself revolutionary. It changed the way people look at the world around them—so much so that "paradigm" and "paradigm shift" have entered general usage. Although Kuhn intended the term "paradigm" to apply to science, scholars have imported it into other academic disciplines. The world of commerce has adopted it, too. Unsurprisingly, "paradigm" now appears on lists of words that have been so overused as to become meaningless.

Although Kuhn intended the book to influence readers interested in the history and philosophy of science, it has also influenced scientists in other fields, teachers of science, academics interested in the sociology of science, and even the general public. Today, radical intellectuals remain attuned to Kuhn's ideas. *Structure* has become a foundational text for people interested in postmodernist* critiques of science. Roughly, these approaches, often employing the theory that science is simply another culturally bound method of producing narratives, are frequently used to demonstrate ways in which historically marginalized groups such as women, people of color, and colonial subjects, have impacted on the development of science. The new discipline of the sociology of scientific knowledge,* meanwhile, maintains that social conditions affect the creation of scientific knowledge—a very Kuhnian view.

While many scientists continue to disagree with Kuhn, they acknowledge the large contribution his work made to science. One of his longstanding critics, the German American philosopher of science Carl Hempel,* acknowledged as much. Paying tribute to Kuhn in 1993, Hempel wrote: "Whatever position your colleagues may take, Tom, I am sure that they all feel a large debt of gratitude to you for your provocative and illuminating ideas."[2]

Over half a century after its publication, *Structure* continues to inspire and provoke readers. Even scholars working in non-scientific fields speak of the "structure of revolutions" as Kuhn defined the term.

---

1. N. M. Swerdlow, "Thomas S. Kuhn, A Biographical Memoir," *National Academy of Science*, (2013), accessed June 29, 2015, http://www.nasonline.org/ publications/biographical-memoirs/memoir-pdfs/kuhn-thomas.pdf.
2. Carl Hempel, "Thomas Kuhn: Colleague and Friend," in *World Changes: Thomas Kuhn and the Nature of Science*, ed. Paul Horwich (Cambridge, MA: MIT Press, 1993), 7–8.

SECTION 1
INFLUENCES

MODULE 1
THE AUTHOR AND THE HISTORICAL CONTEXT

KEY POINTS

- *The Structure of Scientific Revolutions* presents an ongoing challenge to the notion that science can discover the truth about what reality "is."
- Thomas Kuhn's work as a radar technologist and his graduate studies in physics left him dissatisfied with fundamental aspects of how scholars and the public understood science.
- Kuhn's work was shaped by the events of World War II* and the Cold War* (a long period of frequently grave political tension between the United States and its allies and the Soviet Union* and its allies).

Why Read This Text?

Thomas Kuhn's *The Structure of Scientific Revolutions* presents a thorough analysis of how scientific knowledge evolves. Regularly found on lists of "best books of the twentieth century," and one of the most popular non-fiction books of the last half-century, Kuhn's work changed the way scientists approached their task and the way we think about breakthroughs, whether in science or in other spheres of human endeavor. Its key concepts—"paradigm,"* an intellectual model for understanding the world; "scientific revolution,"* the moment when one paradigm gives way to another; and "paradigm shift,"* the state of affairs after a scientific revolution—have entered the popular vocabulary.

Although the idea that changes in human understanding can be revolutionary has entered everyday language,[1] the work's core question—what forces can change the ways scientists think and work?—remains hotly debated today. Previously, the idea of a scientific revolution had been narrowly applied to physics. Kuhn extended the term to apply to all areas of science; his many readers have used it to refer to intellectual endeavor in general.[2]

Kuhn's notion of science as a constantly evolving body of knowledge has affected political culture, both in his native United States and throughout the West. His conclusions demonstrate the provisional and uncertain nature of scientific knowledge, while his historical analysis of the evolution of science gave rise to the discipline of the history of science*; this showed that scientific knowledge, rather than being fixed and certain, actually changed over time.[3]

If science is provisional rather than definitive, however, politicians and public figures can no longer hold up scientific findings as absolute proof of anything.[4]

> "A softer skepticism, one more sympathetic to the aspirations of science, does not renounce the possibility of objective truth, but instead is agnostic about that possibility. Thomas Kuhn is such a skeptic."
>
> ——Mike W. Martin, *Creativity: Ethics and Excellence in Science*

Author's Life

Kuhn was a scientist for much of his life. At school he developed a love for mathematics and physics[5] and went on to Harvard University,* completing his undergraduate and doctoral studies in physics by 1949.[6] Harvard's influential president James Bryant Conant* helped Kuhn obtain professional positions in the history and science departments.[7]

Conant had been an important figure in the development of science during World War II. As administrator of the research and development program known as the Manhattan Project,* he not only helped develop the atomic bomb but convinced US President Harry S. Truman* that its use was inevitable.[8] After the war, Conant returned to the presidency of Harvard and began a project to bring science to a wider audience—both in academia and in the public. He enlisted the support of such senior scholars as the Belgian émigré historian of science and chemist George Sarton* as well as junior scholars such as Kuhn.[9]

In the course of his career Kuhn continued Conant's undertaking, devoting himself to teaching the history and philosophy of science* in ways accessible to both the general public and scholars. In 1961, he became a professor of the history of science at the University of California,* where he wrote *The Structure of Scientific Revolutions.* On leaving in 1964, Kuhn taught at Princeton University* and the equally prestigious Massachusetts Institute of Technology.* He also served for a year as president of the History of Science Society.*[10]

Before the 1960s—that is to say, before the publication of *The*

Structure of Scientific Revolutions—scientists who wished to write about the history of science did so as a sideline to their main jobs.[11] Kuhn's work changed that. He made it possible for scientists to cross over from physics to the history and philosophy of science. In doing so, he changed the world's understanding of science forever.

Author's Background

After receiving his undergraduate degree from Harvard during World War II, Kuhn spent two years in the military working on radar technology. His occupation left him unhappy with how the world outside of academia applied and practiced science. After the war he returned to Harvard for postgraduate study.[12] Although formally a student of physics, he also wanted to look beyond the application of science in the world to the underlying principles of scientific knowledge itself.[13]

In 1945, near the end of World War II, the American government forced Japan's surrender by unleashing a display of overwhelming military force—the detonation of the atomic bomb, which Kuhn's mentor, Conant, had helped to develop. The bomb destroyed two Japanese cities, Hiroshima* and Nagasaki,* and prompted politicians and the general public to ask whether it was morally justifiable to use science and technology in warfare.[14] Whatever view people took, events seized public attention: scientific discoveries had the power to shape human destiny.

The Cold War, which began as World War II ended and lasted until around 1991, also shaped Kuhn's views about science. America's rivalry with the Soviet Union saw scientific and technological progress became topics of public debate; in particular, Kuhn focused on the development of nuclear weapons, and the threat that nuclear war could break out between these two powerful nations. Physics played a central role in these major developments. The Canadian historian of science Ian Hacking* writes that during this time, "everybody knew that physics was where the action was"[15]—including Kuhn. The science of physics and its foundational principles and characteristics were ripe for analysis. Kuhn's whole career can be seen as a response to this challenge.

---

1. Thomas Nickles, "Introduction," in *Thomas Kuhn*, ed. Thomas Nickles (Cambridge: Cambridge University Press, 2003), 1; Ian Hacking, "Introductory Essay," in Thomas S. Kuhn, *The Structure of Scientific Revolutions*, 4th ed. (Chicago, IL: University of Chicago Press, 2012), xxxvii.
2. Kuhn, *Structure*, 32.
3. Christopher Green, "Where Is Kuhn Going?," *American Psychologist* 59, no. 4 (2004): 271–2.
4. Ian Hacking, *The Social Construction of What?* (Cambridge, MA: Harvard University Press, 1999), 12.
5. Alexander Bird, "Thomas Kuhn," *The Stanford Encyclopedia of Philosophy* (Winter 2014 Edition), Edward N. Zalta, ed., http://plato.stanford.edu/ archives/win2012/entries/davidson/.
6. Bird, "Thomas Kuhn."
7. Steve Fuller, *Thomas Kuhn: A Philosophical History for Our Times* (Chicago, IL: University of Chicago Press, 2000), 9–11.
8. Ziauddin Sardar, "Thomas Kuhn and the Science Wars," in *Postmodernism and Big Science*, ed. Richard Appignanesi (Cambridge: Icon Books, 2002), 200.
9. Fuller, *Thomas Kuhn*, 9–11.
10. Bird, "Thomas Kuhn."
11. Jeff Hughes, "Whigs, Prigs and Politics: Problems in the Historiography of Contemporary Science," in *The Historiography of Contemporary Science and Technology*, ed. Thomas Söderqvist (Amsterdam: Harwood Academic Publishers, 1997), 20–1.

12. Bird, "Thomas Kuhn."
13. Marnie Hughes-Warrington, "Thomas Samuel Kuhn," in *Fifty Key Thinkers in History* (London: Routledge, 2003), 188.
14. Thomas C. Reeves, *Twentieth-Century America: A Brief History* (Oxford: Oxford University Press, 2000), 137.
15. Hacking, "Introductory Essay," in Kuhn, *Structure*, ix.

MODULE 2
ACADEMIC CONTEXT

KEY POINTS

- The fields of the history of science* and the philosophy of science* explain the evolving ways in which scientists have arrived at their conclusions.

- In the 1960s, scholars challenged the idea that science provides concrete, accurate results. Supporters of scientific realism* (who argue that a scientist's opinions do not necessarily affect their conclusions) faced off against supporters of scientific constructivism* (who argue that a scientist does not study reality directly—only information collected through experiments).

- Thomas Kuhn led the way in using historical analysis to undermine both logical empiricism* and logical positivism*— approaches to science and philosophy grounded in verifiability,* logic, and rationalism.*

The Work in its Context

Thomas Kuhn's work *The Structure of Scientific Revolutions* falls within the field of the history of science: the study of how scientific knowledge evolves over time, a field pioneered by leading sixteenth-century scientists such as Francis Bacon* in England and Johannes Kepler* in Germany. In their lifetimes, scientists such as Isaac Newton* (a physicist who outlined principles that continue to inform the study of physics today[1]) made significant discoveries. Bacon and Kepler wanted their histories of science to reconstruct a "usable past" showing how the "modern science" of their lifetimes had developed since earlier, "ancient wisdom."[2]

At the outset of the twentieth century, the history of science remained little changed from Bacon and Kepler's day. Twentieth-century historians of science, for instance, saw scientific knowledge as an accumulation of facts and data. As scientific experiments added to this store of facts, they saw science as having a linear trajectory, with all new scientific discoveries building on previous discoveries to produce "progress" in scientific knowledge. In addition, scientists passively observed and recorded their results without injecting their own opinions or prejudices into their findings.[3]

> "Kuhn addressed philosophical questions about reason and evidence via an examination of history ... The logical empiricists made a sharp distinction between questions about the history and psychology of science, on the one hand, and questions about evidence and justification, on the other. Kuhn was deliberately mixing together things that logical empiricists had insisted should be kept apart."
>
> —— Peter Godfrey-Smith, *Theory and Reality: An Introduction to the Philosophy of Science*

Overview of the Field

By the time Kuhn wrote *Structure* in 1962, the history of science had witnessed significant, transformative developments. Previously it had been written by scientists but now historians, even those without formal scientific training, had become interested in the subject as well.[4] The Belgian-born American scientist and historian

George Sarton* did the most to initiate this transition, explaining that "historians of science must know history *and* science ... good intentions are never enough."[5]

Historians studying science drew on an emerging historical sub-discipline: the history of ideas.* Pioneered in the US by the German-born American philosopher and historian Arthur O. Lovejoy,* the history of ideas showed how ideas evolved to mean different things over the centuries. Lovejoy "diligently developed the discipline, framing it as a study of purely cognitive worldviews* and their component 'unit ideas'."*[6]

In other words, Lovejoy argued that people, scientists included, respond to the world by forming mental images—"cognitive worldviews." They use these mental images to connect the collection of ideas they hold in their minds, isolated ideas he called "unit ideas." Any idea about the world, whether of science, gravity or death, for example, would change over time because people would put their unit ideas together in different ways. So, different people, places, and eras would hold different ideas about science, gravity or death—or anything else they constructed from their unit ideas.

Academic Influences

Kuhn's influences came from scholars working in science, in social science, and in the history of science. He also benefited from interactions between these disciplines.

When the scientist and Harvard University president James Bryant Conant* set out to educate the public about science, he enlisted Kuhn's help. In researching public science education, Kuhn noticed a contrast between "modern science" and the "ancient wisdom" Kepler and Bacon had written about.[7] Conant also introduced Kuhn to Sarton—another member of the science education program—and learned about Sarton's innovative ways of thinking about the history of science.[8]

Scholars from the social sciences, psychology, and linguistics inspired Kuhn to think in new ways about how people's minds and language shape the way they see the world.

The French psychologist Jean Piaget's* theories of child development and the experiments of the school of Gestalt psychology* both emphasized that people use a theoretical framework to analyze the world around them. Kuhn accepted this.[9] In other words, he agreed that our thought processes could not be separated from what we know about the world. From linguistics, Kuhn learned about the ways context changes the meaning of words. This reinforced his understanding of Piaget's psychology that each individual had different perspectives.

Kuhn's greatest influence, however, was the Austrian-British philosopher Ludwig Wittgenstein.* Wittgenstein argued that when individuals say something about the world, they are "interpreting," and "when we interpret we form hypotheses, which may prove false."[10] In other words, people may see the world incorrectly even though what they say is true at any given moment.

Wittgenstein was inspired to write his theory of language by the eighteenth-century German philosopher Immanuel Kant.* Kuhn also studied him closely. Kant identified four categories one could use to analyze anything: quantity, quality, relation, and modality. Using these categories, Kant could describe an object in terms both of its own characteristics and by reference to the context in which it existed. Previous philosophers had focused on just one or the other.[11] In other words, Kant's categories allowed him to formulate a way of interpreting how physical objects would behave.

Kuhn also wanted to explain scientific activity with reference to foundational concepts. Kant's categories informed Wittgenstein's twentieth-century theory of how we use language to interpret the world. By extension, Kant's ideas showed Kuhn how our everyday speech shaped our understanding of the world.

Historians of science inspired Kuhn to focus on the role cultural assumptions play in scientific conclusions. Alexandre Koyré,* a French Russian scholar who immigrated to America, resurrected the idea that scientific knowledge progressed because of changes, or revolutions, in the way scientists thought.[12] Kuhn also found thought-provoking the French philosopher of science Hélène Metzger's* work on seventeenth-century chemistry and the German historian Anneliese Maier's* work on medieval science. But maybe the most important source of inspiration to Kuhn was the work on tacit knowledge* (roughly, the kind of knowledge not easily transferred from one person to another by the simple act of writing or speaking) of Hungarian British scholar Michael Polanyi,* who excelled in many disciplines. Polanyi's thinking encouraged Kuhn to investigate the ways in which cultural attitudes, as much as scientific method, shaped and could be detected in scientists' conclusions.[13]

1. A. Rupert Hall, *The Revolution in Science 1500–1750* (London: Longman, 1983), 134, 143.
2. Hall, *Revolution in Science 1500–1750*, 18.
3. Thomas S. Kuhn, *The Structure of Scientific Revolutions*, 4th ed. (Chicago, IL: University of Chicago Press, 2012), 31.
4. Jeff Hughes, "Whigs, Prigs and Politics: Problems in the Historiography of Contemporary Science," in *The Historiography of Contemporary Science and Technology*, ed. Thomas Söderqvist (Amsterdam: Harwood Academic Publishers, 1997), 20.
5. George Sarton, *A Guide to the History of Science* (Waltham, MA: Chronica Botanica Co, 1952), ix–x.
6. Andrew Jewett, *Science, Democracy and the American University: From the Civil War to the Cold War* (Cambridge: University of Cambridge Press, 2012), 257.
7. Thomas Kuhn, *The Road Since Structure: Philosophical Essays, 1970—1993 with an Autobiographical Interview*, ed. James Conant and John Haugeland (Chicago, IL: Chicago University Press, 2000), 16.
8. Steve Fuller, *Thomas Kuhn: A Philosophical History for Our Times* (Chicago, IL: University of Chicago Press, 2000), 9–11.
9. Kuhn, *The Structure of Scientific Revolutions*, "Preface," xi.
10. Ludwig Wittgenstein, *Philosophical Investigations*, 3rd ed. (Oxford: Blackwell, 2003), 181.
11. Michela Massimi, "Philosophy and the Sciences After Kant," in *Conceptions of Philosophy*, ed. Anthony O'Hear (Cambridge: Cambridge University Press, 2000), 282.
12. Paul Hoyningen-Huene, *Reconstructing Scientific Revolutions: Thomas S. Kuhn's Philosophy of Science* (Chicago, IL: University of Chicago Press, 1993), xix.
13. Alexander Bird, *Thomas Kuhn* (Chesham: Acumen, 2000), 14–20.

MODULE 3
THE PROBLEM

KEY POINTS

- Historians of science question whether scientists work objectively, studying reality unaffected by their personal views or professional positions.
- Logical empiricists* argue that scientists verify all their discoveries by reference to scientific data. But historians of science contend that the concept of evidence itself has changed over time, making verification relative.
- Kuhn combines these positions, using verification and relativity to produce a new understanding of scientific knowledge.

Core Question

In *The Structure of Scientific Revolutions*, Thomas Kuhn sought "to clarify and deepen an understanding of *contemporary* scientific methods or concepts displaying their evolution."[1] In other words, he reframed the central question asked by historians of science since the discipline emerged in the sixteenth century: how do scientists find out about the real world?

This central question breaks down into two sets of questions: those about *scientists* and those about the *nature of reality itself*. Historians asked if scientists could escape their personal and cultural prejudices when they evaluated data. And they asked if reality exists independently of the mind of the scientist responsible for evaluating it.

These questions uncover a foundational problem that continues to divide historians and scientists.² On one side, "realists"* argue that scientists have direct access to the real world of objects and matter, and that they analyze, experiment on and describe it. In this way, realists believe that scientists can uncover the truth without allowing their personal opinions to interfere with their research.³ The world is as it is and it is plain for all scientists to see for themselves. Other scholars—"constructivists"*—hold the opposite view. They believe that scientists construct their own picture of reality. Scientists cannot help allowing their own ideas, perceptions, and intuitions to shape that picture. And their own research and their standing in the scientific community,* along with a host of intellectual and social variables, also shape their ideas of reality.⁴

> "There is an increasing tendency, now, to construe knowledge as a construct of historical experience and to challenge any principled disjunction between objectivity in natural sciences and in the human sciences and practical life. To favor these themes is, effectively, to deny any privilege or hierarchical order of knowledge favoring the natural sciences."
> ——Joseph Margolis, "Objectivity as a Problem"

The Participants

Before Kuhn published *The Structure of Scientific Revolutions*, realist and constructivist positions had their own representatives. Historians of science stood with the constructivists, and logical positivists or empiricists* (whose approach "is often summarized by saying that the only source of knowledge is experience")took the realist position. ⁵

Before 1960, logical empiricism—a school in the philosophy of science* with roots in a tradition begun by the seventeenth-century English philosopher John Locke* and continued by the likes of the Scottish philosopher David Hume* in the eighteenth century and the German philosopher Ernst Mach* in the nineteenth—seemed the stronger argument. Around the time of World War I,* the German philosophers Rudolf Carnap,* Otto Neurath,* and Carl Hempel* perceived a decline in standards of philosophy. Their work—which came to be known as logical empiricism—served as a response to this.⁶ Logical empiricism updated the classical empiricist tradition founded by Locke by adding two central ideas: the analytic-synthetic distinction,* and the idea of the verifiability theory of meaning.*

The German philosopher Immanuel Kant* introduced the idea of analytic-synthetic distinction in the seventeenth century. An analytic proposition* would be a statement such as "all bachelors are unmarried." The definition of a bachelor—an unmarried man—is contained in the proposition itself. By contrast, a synthetic proposition,* such as "all bachelors are unhappy," does not contain its definition. "Unhappy" is not part of the definition of "bachelor" and the statement may be true for some bachelors, but not for others. Logical empiricists paired this centuries-old concept with a newer one—the verificationist principle* which maintained that only two kinds of statements hold any real meaning: those that are logically necessary and those that can be verified by experiment.

According to the idea of verification, another scientist can test what the first scientist comes to understand about the whole only by repeating her experiments, ensuring the validity of her analysis and synthesis.[7]

Combining analytic-synthetic distinction and the verifiability theory, logical empiricists would say that a scientist reaches an understanding of the *whole* perceived from experience of the world by using the rational process of analyzing and thinking about data relating to *parts of that whole*.[8]

Logical empiricists held that these basic concepts applied to all science at any time, and, as a result, that science continually accumulated new knowledge and discoveries as it had for the sixteenth-century originators of the history of science*: the English philosopher Francis Bacon* and the German astronomer Johannes Kepler.*

Historians, by contrast, occupied a constructivist position. As they saw it, scientists operated according to principles unique to their own age. Looking across the centuries, they saw science developing in an irregular, unpredictable manner. They insisted that to recreate an accurate history of science, one needed to "contextualize" scientific activity.

In the mid-twentieth century, historians of science pointed to the work of the late nineteenth-century French philosopher and historian Pierre Duhem* who argued that scholars need a "rationally reconstructed [history of science in which] only what was philosophically relevant was included."[9] Historians, in other words, should look at what scientists believed and how they acted on it. Historians of science should investigate this if they wanted to find out how science had evolved over time, instead of trying to apply general philosophical concepts as logical empiricists did.

The Contemporary Debate

By the 1960s, when Kuhn wrote *The Structure of Scientific Revolutions*, this debate between logical empiricists and historians had become entrenched. Overlaps developed between the two sides, and Kuhn took elements from each.

Both sides relied on a particular understanding of the role of language in creating knowledge. The logical empiricists had derived their theory of verifiability and the distinction between analysis and synthesis from the work of linguists. In addition to Kant, they drew on the philosophy of the twentieth-century philosopher Ludwig Wittgenstein,* both of whom, working centuries apart, emphasized underlying mental operations as key to understanding how individuals acquire knowledge.[10] Kuhn's interest in Kant confirmed his connection to the logical positivists; all wanted to ground their universal understanding of science on basic principles with logical foundations.[11]

Although Kuhn seldom associated himself publicly with the logical empiricists, he did associate himself with the contextual interpretation of science offered by constructivists. Indeed, he maintained that he conceived of *The Structure of Scientific Revolutions* as a stand against the logical empiricism that dominated science textbooks of the 1950s.[12] He claimed he did so by proposing that science operated in disjointed cycles—so scientific knowledge developed unevenly, an idea that stood in stark contrast to assumptions of the logical empiricists. He called his view a "historiographic revolution in the study of science" that borrowed extensively from Duhem, Alexandre Koyré* and others to promote a historical rather than a philosophical understanding of science.[13]

The history of science that emerged from the work of Duhem and the Franco-Russian science historian Alexandre Koyré, who coined the phrase "scientific revolution," and who Kuhn praised throughout *Structure*, led Kuhn down the path of historical interpretation. In Kuhn's view, the history of science had languished as a peripheral subject in both history and science departments. But Koyré's work gave it scholarly credibility.[14] Kuhn argues that Koyré paved the way for a new approach to writing history texts that can "suggest the possibility of a new image of science."[15] In other words, Kuhn believed that the findings of historical studies could revolutionize the understanding of how science works. A lively contemporary debate raged and, with the publication of *Structure*, Kuhn occupied a central position within it.

---

1. Thomas S. Kuhn, "The History of Science," in *The Essential Tension: Selected Studies in Scientific Tradition and Change* (Chicago, IL: University of Chicago Press, 1977), 107.
2. Peter Godfrey-Smith, *An Introduction to the Philosophy of Science: Theory and Reality* (Chicago, IL: University of Chicago Press, 2003), 5.
3. Ian Hacking, *The Social Construction of What?* (Cambridge, MA: Harvard University Press, 1999), 68.
4. Godfrey-Smith, *Theory and Reality*, 6.
5. Godfrey-Smith, *Theory and Reality*, 228.
6. Godfrey-Smith, *Theory and Reality*, 28.
7. Godfrey-Smith, *Theory and Reality*, 28.

8. Godfrey-Smith, *Theory and Reality*, 27.
9. R. N. D. Martin, *Pierre Duhem: Philosophy and History in the Work of a Believing Physicist* (La Salle, IL: Open Court, 1991), 139.
10. Thomas S. Kuhn, "Afterwords," *World Changes: Thomas Kuhn and the Nature of Science*, ed. Paul Horwich (Cambridge, MA: MIT Press, 1993).
11. Thomas S. Kuhn, *The Road Since Structure: Philosophical Essays, 1970–1993 with an Autobiographical Interview*, ed. James Conant and John Haugeland (Chicago, IL: Chicago University Press, 2000), 264.
12. Thomas S. Kuhn, *The Structure of Scientific Revolutions*, 4th ed. (Chicago, IL: University of Chicago Press, 2012), 2.
13. Kuhn, *Structure*, 3.
14. Kuhn, *Structure*, 3. In a 1971 piece he talked about Koyré as being his "maître" above all other historians; Thomas S. Kuhn, "Concepts of Cause in the Development of Physics," in *The Essential Tension*, 21 (this is the published translation of the originally French piece).
15. Kuhn, *Structure*, 3.

MODULE 4
THE AUTHOR'S CONTRIBUTION

KEY POINTS

* Thomas Kuhn argued that science followed this course: Pre-consensus science* Normal science* Crisis science* Extraordinary science* Scientific revolution.*
* *The Structure of Scientific Revolutions* overturned the dominant view of linear, cumulative progress in science by proposing this cyclical alternative.
* Although the cyclical model developed the old concept of a scientific revolution, its originality lay in the challenge it posed to objectivity (that is, the assumption that a scientist was capable of examining "reality" without his or her own world view intervening).

Author's Aims

In *The Structure of Scientific Revolutions*, Thomas Kuhn aimed to demonstrate that scientific knowledge progresses in cycles—a theory he began to develop during his doctoral studies when, as he wrote, his "exposure to out-of-date scientific theory and practice radically undermined some of my basic conceptions about the nature of science."[1]

This cyclical understanding began with an acknowledgement that historians of science had themselves framed their histories with concepts that changed over time.[2] If the analysis of history can be cyclical, then so can knowledge itself.

Kuhn demonstrated the cycles moving from "pre-consensus science"—when rival schools of thought addressed the same object from different perspectives, reaching different conclusions[3]—through stages to revolution and back again. Pre-consensus science, for Kuhn, is "something less than a science"[4] because no shared opinions exist and any idea can be tried. He cites as an example the different interpretations held by scientists in the seventeenth century, before the work of the English physicist Isaac Newton,* of how light waves behave—a domain of physics known as optics.[5]

Out of this "pre-consensus science" grows "normal science" in which one school of thought wins the allegiance of all scientists, who then work to modify agreed concepts by consensus within accepted rules of scientific practice,[6] accumulating knowledge in a linear manner.[7]

This gives way to crisis and extraordinary science and eventually revolution. A crisis in science occurs when scientists, faced with inexplicable phenomena that arise during normal science, can no longer explain the world. What Kuhn calls "science in a crisis state"[8] reconsiders the conventions and agreed concepts that direct normal science. When crises occur, several paradigms* compete for the acceptance of the scientific community.* These paradigms, in Kuhn's view, are incommensurable*—that is, they are worldviews that explain reality so differently that they cannot coexist.

Kuhn argues that it is scientists' personalities and preferences that ultimately explain the decisions scientific communities make about which paradigm will be accepted. If paradigms explain the world differently, but each is plausible according to its own argument, then choices are made for reasons that are not scientific but personal.[9]

By disrupting the linear process of normal science, these crises make scientific development uneven. They do not always end in revolution. Most often, they find a resolution. Perhaps a solution emerges from the "normal science" camp. Or perhaps the scientific community agrees to set the problem aside for future generations. But when such solutions fail, a revolution may take place. After the revolution, science reverts back to "normal science" and the whole process repeats itself.

If a scientific field reaches the revolutionary stage, the effects are dramatic. Revolutions create new paradigms, which require the complete "reconstruction" of a field of science.[10] By the end of the revolution, the profession itself will have changed its views, methods, and goals.[11]

Alluding to the psychological theory of "gestalt"*—which assumes that we build a worldview by mentally ordering our experience in such a way that some individual, new, experience has the capacity to radically alter our understanding of what is "real" altogether—Kuhn compares the change to a visual gestalt whereby an animal previously seen as a bird is now seen as an antelope.[12] The revolutionary paradigm shift* produces an equal displacement in scientists' conceptual framework.[13]

> "The origin of *The Structure of Scientific Revolutions* explains why the questions there raised are similar to those asked by recent converts to the history of science who have been trained as scientists and by historians strongly influenced by sociology; indeed Kuhn regards this book as a contribution to the sociology of knowledge."
>
> —— Marie Boas Hall, review of *The Structure of Scientific Revolutions*

Approach

Kuhn seeks to look at science in its context, as scientists themselves have interpreted it. He says he is trying to "display the historical integrity of [a] science in its own time" by considering what is going on in the minds of a group of scientists practicing "some particular scientific specialty during some particular period."[14]

Although Kuhn attaches great importance to history as a starting point for analyzing scientific activity, this is not the end of his approach. From a historical understanding of how scientists work, he maintains that he can deduce a series of concepts explaining how and why they worked, and how they continue to work.[15]

This approach contrasts with that of logical empiricism,* according to which a scientific experiment that cannot be verified proves very little about reality. Consequently, the relationship between Kuhn's thought and that of logical empiricism, remains unclear, and Kuhn himself often avoided the comparison.[16]

Kuhn drew heavily on the work of early twentieth-century historians such as the Franco-Russian Alexandre Koyré* (coiner of the phrase "scientific revolution") and the French historian Pierre Duhem.* These scholars sought to show the true character of science by studying episodes in the history of science* in detail to see how scientists had actually worked.

Where logical empiricists explained historical science by turning first to philosophical concepts, Kuhn effectively did the reverse, putting history first to reveal philosophical concepts:

"Kuhn's work seemed to show how interesting it is to connect philosophical questions about science with questions about the history of science."[17]

Contribution in Context

The originality of *The Structure of Scientific Revolutions* lies in the fact that it builds on existing frameworks for understanding the history of science—notably the idea that science occupies a special place in human intellectual enterprise—while destabilizing those same frameworks by arguing that scientific advances are cyclical and that scientists have a large say in defining "reality." When leading logical empiricists sought to assemble a definitive collection of statements about the nature, function, and purpose of science,[18] Kuhn offered to let them include *The Structure of Scientific Revolutions* in their project, the *Encyclopedia of Unified Science*.

Another highly original aspect of Kuhn's work was his subversion of existing concepts used both by logical empiricists and historians. Kuhn shocked the scientific establishment by showing that apparently settled ideas contained within them the seeds of radical reinterpretation—the polar opposite of the dream of logical empiricists to assemble a definitive collection of statements about science.

In 1985, the American philosopher Arthur Danto* said that Kuhn ensured the onset of "post-empiricism," writing that there "really was a unity of science, in the sense that all of science was brought under history rather than, as before, history having been brought under science construed on a model of physics."[19]

Kuhn's originality lay, then, in creating unity by showing the inherent disunity of science, and so moving science beyond logical empiricism.

---

1. Thomas S. Kuhn, *The Structure of Scientific Revolutions*, 4th ed. (Chicago, IL: University of Chicago Press, 2012), v.
2. Kuhn, *Structure*, 2, 7; Kuhn, *The Essential Tension: Selected Studies in Scientific Tradition and Change* (Chicago, IL: University of Chicago Press, 1977), xi.
3. Kuhn, *Structure*, 4, 12–3, 47–8, 61–2, 178–9.
4. Kuhn, *Structure*, 13–8.
5. Kuhn, *Structure*, 16.
6. Kuhn, *Structure*, 17–9, 178.
7. Kuhn, *Structure*, 52–3.
8. Kuhn, *Structure*, 82–7, 101, 154.
9. Kuhn, *Structure*, 95.
10. Kuhn, *Structure*, 85.
11. Kuhn, *Structure*, 85.
12. Kuhn, *Structure*, 85.
13. Kuhn, *Structure*, 102.
14. Kuhn, *Structure*, 3.
15. Kuhn, *Structure*, 3.
16. Joseph Rouse, "Kuhn's Philosophy of Scientific Practice," in *Thomas Kuhn*, ed. Thomas Nickles (Cambridge: Cambridge University Press, 2003), 101.
17. Peter Godfrey-Smith, *An Introduction to the Philosophy of Science: Theory and Reality* (Chicago, IL: University of Chicago Press, 2003), 78.
18. Peter Novick, *The Noble Dream: The "Objectivity Question" and the American Historical Profession* (Cambridge: Cambridge University Press, 1988), 526.
19. Arthur Danto, *Narration and Knowledge* (New York: Columbia University Press, 1985), xi–ii.

SECTION 2
IDEAS

MODULE 5
MAIN IDEAS

KEY POINTS
- The main themes of *The Structure of Scientific Revolutions* are scientific revolution,* paradigm,* and incommensurability.*
- Kuhn argues that when scientists challenge and overturn paradigms, they drive scientific progress from normal science* (that is, science practiced according to the prevailing worldview) through crisis* and revolutionary science* back to a new normal science.
- The argument first appears—as a briefer analysis—in two volumes of the *International Encyclopedia of Science*.

Key Themes

In *The Structure of Scientific Revolutions*, Thomas Kuhn constructs his original understanding of science around three key themes:
- The idea of a scientific revolution
- The concept of a paradigm
- The concept of incommensurability

As Kuhn sees it, *scientific revolution* occurs when momentous breakthroughs propel scientific progress. These breakthroughs permanently change the way scientists understand the world.

Kuhn used the word *paradigm* to refer to the core concepts scientists worked with to deepen their knowledge of the world. Kuhn's understanding of science anchors itself firmly around this core idea. He defines it as referring to any "accepted model or pattern" of scientific conduct.[1]

The concept of *incommensurability*, closely related to paradigm, forms the last of Kuhn's core ideas. Incommensurability describes the relationship between paradigms during periods of extraordinary science,* when anomalies* (that is, observations that do not "fit" into the accepted model) cast doubt over the consensus paradigm.* In moments of crisis, when two or more paradigms compete to explain reality differently, the paradigms themselves become incommensurable—that is, scientists using different paradigms cannot communicate with one another. Even if they can, they cannot understand each others' findings because they are using the different standards of evidence or different concepts required by their preferred paradigm. This incommensurability and the chaos that results continues until a new consensus paradigm is reached.[2]

Together, these three ideas enable Kuhn to make his argument for scientific progress. He sees it as cyclical and also as determined by the community* of scientists, since scientists, by their actions, both accept and bring down paradigms.

Kuhn legitimates his ideas by basing his concepts on historical examples of scientific breakthroughs. In considering past scientific advances, he shows how his general concepts can explain and help us understand specific discoveries. He intended his work to definitively discredit the linear, cumulative understanding of science that characterizes the logical empiricist* approach to the philosophy of science.*

> *"Conceived as a set of instruments for solving technical puzzles in selected areas, science clearly gains in precision and scope with the passage of time. As an instrument, science undoubtedly does progress."*
> ——Thomas Kuhn, *The Structure of Scientific Revolutions*

Exploring the Ideas

The key to Kuhn's historical reading of scientific progress lies in the notion of a scientific revolution. His argument is that "each scientific revolution alters the historical perspective of the community [of scientists] that experiences it" with the result that it "should affect the structure of post-revolutionary textbooks and research publications."[3] In other words, every time a scientific breakthrough occurs, scientists use their newfound knowledge to rewrite science. In Kuhn's view, scientific knowledge does not accumulate. Breakthroughs do not add to existing knowledge, they create a "world change" in which scientists see reality differently.[4]

In any scientific revolution, scientists must establish a suitable paradigm to explain what they study. A paradigm emerges to explain and then guide scientists' work on "a few problems that the group of practitioners has come to recognize as acute."[5] According to the German philosopher Paul Hoyningen-Huene,* Kuhn's chief interpreter and defender, paradigms gain acceptance through a two-part process. First, scientists agree that a certain situation presents a scientific problem. Then they agree on a paradigm that provides a scientifically acceptable solution.[6] Paradigms both explain scientific problems and guide scientific practice.[7] Although Kuhn describes them as relatively "rigid," being part of normal science which is itself characterized by rigidity,[8] they are not so inflexible as to cause scientists to end their "puzzle-solving" research.[9]

Scientists will continue to research and encounter new problems. Indeed, paradigms must be relatively open-ended in that they can provide the direction of future research. They help scientists determine which facts are worth gathering and analyzing.[10] Once scientists encounter a problem the paradigm cannot explain, then the concept of incommensurability (that is, the impossibility of reconciling two opposing paradigms) explains how, during a period of crisis science, and even a scientific revolution itself, potential replacement consensus paradigms are considered and discarded.

As we have seen, for Kuhn a paradigm is, fundamentally, "an accepted model or pattern."[11] Hence social acceptance in the community of scientists is essential to the establishment of a paradigm. Kuhn sees a fundamental difference between a paradigm and a rule. For him, rules in the scientific community* are secondary and reductive—and consequently not as productive as paradigms. Rules are also too rigid for scientific enterprise. In conceptualizing the idea of paradigm, Kuhn appeals to the concept of "tacit knowledge"* developed by the British Hungarian philosopher of science Michael Polanyi.* Tacit knowledge is "knowledge that is acquired through practice and ... cannot be articulated explicitly"—in other words, the unspoken guidelines scientists accept.[12]

Kuhn believes scientific paradigms have their own lifespan. They begin by solving a particularly acute problem, and end by failing to answer a subsequent anomaly of similar importance. Yet paradigms do not simply fade away; they resist change until a new paradigm takes over. The process of replacing one paradigm with another occurs in what he calls a scientific revolution—the total renewal of an entire scientific field.

This leads Kuhn to his associated thesis: that of incommensurability. Incommensurability describes the relationship between paradigms. In Kuhn's view, no more than one paradigm can ever exist in a particular field of scientific study at any given time. Paradigms hold absolute authority, dictating the entire landscape of a scientific discipline. When a paradigm takes over, "the profession will have changed its view of the field, its methods, and its goals."[13] Once scientists adopt a new paradigm, the entire "conceptual network" of the world changes in that discipline; scientists operate in a "new world."[14] Kuhn compares the competition between paradigms to the battle between revolutionary factions; the winning party essentially wins all.

To explain Kuhn's views on incommensurability, imagine that "A" represents the "world" and "B" represents the human observer.

Kuhn claims that scientists adopt paradigms before they begin the process of scientific observation. So when the observer B looks at the world A, she or he sees the shape of that world through a paradigm that she or he has already adopted. If B adopts paradigm 1, the act of observation will produce world A1. But if B adopts paradigm 2, observation will render world A2. Because the interpretation begins even before the act of observation, however, the paradigm does not determine the existence of either world. The reality of the worlds is independent of the observer.

Many readers understand incommensurability to mean a radical, complete and instantaneous "world change."[15] That understanding caused critics such as the Australian philosopher Howard Sankey* to accuse Kuhn of being suspect to "some form of idealism."*[16] Idealism holds that the human mind plays a decisive role in shaping how people see the world—people do not have direct access to reality. But this does not describe Kuhn's position:"though the world does not change with a change of paradigm," he writes, "the scientist afterward works in a different world."[17]

In statements like this, Kuhn explains that he does not mean that reality depends on the observer. He means that the sensory perception of reality depends on a previously established paradigm. In his view, science is precisely the ability to fit both theory and perception to an objective,"real" nature.[18]

Nevertheless, Kuhn appears deliberately ambiguous. He recognizes that there is a "real" world; it is simply so remote from human knowledge as to be almost inconsequential.[19] As an example, he offers the revolution in chemistry sparked by the work of the eighteenth-century French chemist Antoine Lavoisier,* the discoverer of oxygen, making the radical observation that "in the absence of some recourse to that hypothetical fixed nature that he 'saw differently,' the principle of economy will urge us to say that after discovering oxygen Lavoisier worked in a different world."[20]

For Sankey, however, these statements signify that the "real" world is "dispensable" or "irrelevant" for Kuhn, as it is essentially unattainable by scientific perception or belief.[21] Kuhn's position rests (Sankey argues) on the idea that scientists cannot access the real world and that the scientific community to which they and their paradigms belong have molded their thought.

The German philosopher Paul Hoyningen-Huene proposes a more nuanced interpretation, differentiating between the world-in-itself (the "real" world) and the phenomenal world (the world of human experience).[22] While the "real world" is independent of science, the phenomenal world, shaped by paradigms, is not—a view derived from the German philosopher Immanuel Kant,* whom Kuhn acknowledged had influenced his thought.[23]

Although Kuhn endorsed Hoyningen-Huene's Kantian interpretation, the British philosopher Alexander Bird* remained skeptical. For him, in pegging his work to Kant's, Kuhn was attempting to give "his earlier thought a (particular species of) philosophical sophistication that it did not really have."[24]

Language and Expression

Kuhn uses the words "paradigm" and "incommensurability" in different ways throughout *The Structure of Scientific Revolutions*, acknowledging that "the concept of paradigm will often substitute for a variety of familiar notions."[25]

The philosopher of science Margaret Masterman* identified 21 meanings of the term "paradigm" in *Structure of Scientific Revolutions*[26]—a linguistic fuzziness representing both a strength and a weakness. On the one hand, an idea as vague as Kuhn's paradigm is difficult for philosophers of science to accept. On the other, Kuhn's ability to condense many connected concepts into a single word would have a major impact on academia and popular culture.

Similarly, Hoyningen-Huene observes that Kuhn uses the term "incommensurability" in two different ways. Initially, it appears in Kuhn's discussion of the way problems and standards change as science goes from a pre-revolutionary to a post-revolutionary mode.[27] But when he returns to the concept later in the work, Kuhn uses the term in a much more radical way, saying: "[The] scientist's perception of his environment must be re-educated ... the world of his research will seem, here and there, incommensurable with the one he had inhabited before."[28]

Kuhn compares this phenomenon with a "gestalt switch,"* a term derived from the Gestalt school of psychology, according to which some new, individual piece of perceptual information can lead to a fundamental transformation in our understanding of what it is we are perceiving. He also cites the revolutionary way the scientific world changed after Nicolaus Copernicus* unveiled his model of the universe in the sixteenth century; "after Copernicus," he writes, "astronomers lived in a different world."[29]

---

1. Thomas S. Kuhn, *The Structure of Scientific Revolutions*, 4th ed. (Chicago, IL: University of Chicago Press, 2012), 23.

2. Kuhn, *Structure*, 85.
3. Kuhn, *Structure*, viii.
4. Kuhn, *Structure*, 19.
5. Kuhn, *Structure*, 24.
6. Paul Hoyningen-Huene, *Reconstructing Scientific Revolutions: Thomas S. Kuhn's Philosophy of Science* (Chicago, IL: University of Chicago Press, 1993), 134–5.
7. Kuhn, *Structure*, 46.
8. Kuhn, *Structure*, 19, 49, 64.
9. Kuhn, *Structure of Scientific Revolutions*, 49.
10. Kuhn, *Structure*, 25–6, 48.
11. Kuhn, *Structure*, 23.
12. Kuhn, *Structure*, 44–5 n.1.
13. Kuhn, *Structure*, 85.
14. Kuhn, *Structure*, 102.
15. Hanne Andersen, Peter Barker and Xiang Chen, *The Cognitive Structure of Scientific Revolutions* (Cambridge: Cambridge University Press, 2006), 106.
16. Howard Sankey, "Kuhn's Changing Concept of Incommensurability," *British Journal for the Philosophy of Science* 44, no. 4 (1993).
17. Kuhn, *Structure of Scientific Revolutions*, 121.
18. Kuhn, *Structure*, 134: "it is hard to make nature fit a paradigm."
19. Kuhn, *Structure*, 111.
20. Kuhn, *Structure*, 118.
21. Sankey, "Kuhn's Changing Concept of Incommensurability," 764.
22. Hoyningen-Huene, *Reconstructing Scientific Revolutions*, 239.
23. Thomas S. Kuhn, *The Road Since Structure: Philosophical Essays, 1970–1993 with an Autobiographical Interview*, ed. James Conant and John Haugeland (Chicago, IL: Chicago University Press), 264.
24. Alexander Bird, "The Structure of Scientific Revolutions and its Significance: An Essay Review of the Fiftieth Anniversary Edition," *The British Journal for the Philosophy of Science* 63, no. 4 (2012): 869.
25. Kuhn, *Structure*, 11.
26. Margaret Masterman, "The Nature of a Paradigm," in *Criticism and the Growth of Knowledge*, ed. Imre Lakatos and A. Musgrave (Cambridge: Cambridge University Press, 1970), 59–89.
27. Kuhn, *Structure*, 103.
28. Kuhn, *Structure*, 112.
29. Kuhn, *Structure*, 117.

MODULE 6
SECONDARY IDEAS

KEY POINTS

- *The Structure of Scientific Revolutions* made valuable contributions to the sociology* of scientific communities* (that is, the study of the world of science in its social aspect), the concept of relativism* (roughly, the belief that "final" answers are impossible in science), and the plurality of natural sciences.
- These ideas derive from Kuhn's overall argument about the cyclical structure of scientific progress.
- Although it is sometimes overlooked, relativism has had the widest impact on social scientists in general. The work also changed the way historians of science think about the sociology of the scientific community and the plurality of the natural sciences.

Other Ideas

The main ideas in Thomas Kuhn's *The Structure of Scientific Revolutions* involve scientific revolution,* paradigm* and incommensurability*—compelling ideas that have produced fascinating developments of the history of science.* In addition the book also contains three important secondary ideas, each enhancing Kuhn's overriding argument about the cyclical nature of scientific advances.

First, Kuhn develops a particular idea of the relationship between the scientist and the scientific community. He looks at science as a "social enterprise," affected by the characteristics of its practitioners and "external social, economic and intellectual conditions."[1] This represents an unusual development in work on the philosophy of science.*

In Kuhn's view, science is unlike any other human activity. In this, he agrees with those who subscribe to the school of thought of logical empiricism.* But where the logical empiricists believe science to be special because it is a uniquely rational accumulation of knowledge, Kuhn believes that we arrive at knowledge in cycles, and that scientists do not merely passively process data about reality. For Kuhn, scientists have a say in what is "real" or not, according to the paradigm they actively choose.

This idea was a challenge to the logical positivism of the 1960s, notably its assumptions about rationality. But it ran the risk of being a self-contradictory argument. How could science retain the special position in human endeavor that Kuhn argued it held if it was neither objective nor a unique, linear accumulation of human knowledge? Had Kuhn tried to say that scientific activity was as subjective as any other intellectual pursuit, this contradiction would not have been an issue.

Second, Kuhn destabilizes the notion of truth in science. His argument about paradigms implies that scientists cannot reach definitive conclusions, since the paradigm is formed and accepted by a scientific community that does not have direct access to reality. This leads to him making remarks, at the end of the work, about "salvaging the notion of 'truth'."[2]

Third, Kuhn examines the pluralism of the sciences. In his view, different branches of the sciences operate with their own paradigms and have their own cyclical developments—a notion that strikes down the image of science as a unified, monolithic enterprise whose foundational principles never vary. As Kuhn sees it, each branch of scientific inquiry—from biology, chemistry, and physics to more specialized sub-disciplines—has its own communities with their own paradigms; scientists and science vary by field and object of research.

> *"Taken as a group or in groups, practitioners of the developed sciences are, I have argued, fundamentally puzzle-solvers ... Like any other value, puzzle-solving ability proves equivocal in application. Two men who share it may nevertheless differ in the judgments they draw from its use."*
>
> ——Thomas Kuhn, *The Structure of Scientific Revolutions*

Exploring the Ideas

Kuhn sees the scientific community as a closed unit, separate from society, pursuing its own agenda. Once scientists debate and adopt a paradigm they begin to use an "esoteric* language"—an exotic vocabulary unintelligible to the general public.[3]

Scientists could choose to pursue noble goals that would benefit the "welfare of mankind."[4] Instead, Kuhn sees them looking inward, intent only on improving and cultivating their reputations. Kuhn explains that "unlike the engineer, and many doctors, and most theologians, the scientist need not choose problems because they urgently need solution;"[5] instead, "many of the greatest scientific minds have devoted all of their professional attention to demanding puzzles of this sort. On most occasions any field of specialization offers nothing else to do, a fact that makes it no less fascinating to the proper sort of addict."[6]

In this way, Kuhn blew apart the old understanding of scientists as a venerable community serving mankind. It opened the scientific sphere to the attention of sociologists of science. By arguing that scientists and non-scientists were the same, he implied that their social interactions demanded the same scrutiny.[7] In making this argument, he insisted nevertheless that science itself differed from other human endeavors.

The absence of consensual agreement—the paradigm—voids science: "Once a first paradigm through which to view nature has been found, there is no such thing as research in the absence of any paradigm. To reject one paradigm without simultaneously substituting another is to reject science itself."[8] Although social scientists later deployed this notion in other fields, this was not Kuhn's purpose. For him, the paradigm was a definitive attribute of science.

Kuhn's destabilization of notions of truth also related to the paradigm concept. In his view, no paradigm can ever explain reality as it truly exists. Eventually every paradigms falls, making way for a new one, leaving nothing of any permanence in scientific conclusions but creating continual progress.[9]

This concept gave rise to the idea of relativism, which holds that scientific knowledge has no universal validity. Instead, scientific knowledge relates to the state of scientific development at any given time. As Kuhn wrote, "to be admirably successful is never, for a scientific theory, to be completely successful."[10]

In Kuhn's paradoxical narrative, the more precise the paradigm is, the more likely it is to fail. Anomalies* become much more evident against a background of increased precision. Such "technical breakdown" induces crises in normal science.*[11] We have, therefore, to "relinquish the notion, explicit or implicit, that changes of paradigm carry scientists and those who learn from them closer and closer to the truth."[12]

During periods of normal science, fundamental progress occurs as research makes numerous discoveries using the consensus paradigm.* Kuhn believes that this is a natural consequence of the entire scientific community being in consensus. Thus, "the reception of a common paradigm [frees] the scientific community from the need constantly to re-examine its first principles." Its members can therefore focus their attention "upon the subtlest and most esoteric of the phenomena that concerns it."[13] In this respect, progress can be defined as an increase in precision that improves on the original paradigm.

Yet, surprisingly, there is a sense in which progress across paradigms is possible. As Kuhn sees it, a paradigm never completely erases the previous paradigm's findings. On the contrary, new paradigms "usually preserve a great deal of the most concrete parts of past achievement and they always permit additional concrete problem-solutions besides."[14] Implicitly, a sense of progress arises as new developments incorporate older ones. As Kuhn puts it, scientific development is an enterprise characterized "by an increasingly detailed and refined understanding of nature."[15]

To clarify his point, Kuhn uses the analogy of Darwinian*

selection. He describes the English naturalist Charles Darwin's* biological work in the nineteenth century as fundamentally a movement away from teleology* (the idea that development occurs towards a pre-decided goal) to an understanding of the natural world as the result of unplanned progress that can be measured only against its earlier incarnations. The result is increased "articulation and specialization" but no clear goal.[16] Remarkably, Kuhn never set aside this analogy with Darwinian evolution, even in his late works.

Kuhn's relativism also links to the process by which scientists choose paradigms. Once science is in crisis and competing theories emerge, the success of a paradigm "can never be unequivocally settled by logic and experiment alone."[17] Instead, it relies on such factors as persuasion, power struggles amongst junior and senior scientists, individual scientists' support from university administration, and the availability of funding and financial support for research. Kuhn makes it clear in *The Structure of Scientific Revolutions* that paradigms do not succeed because they are *true* but ultimately because the dominant factions within the scientific community choose one paradigm over another.

Alluding to the novel *1984*, the English writer George Orwell's* vision of a future in which individuality has been erased, Kuhn writes that a newcomer to a scientific community is "like the typical character of ... *1984*, the victim of a history rewritten by the powers that be."[18] He or she is not trained to question the paradigm, but to accept it; in this sense, in scientific communities education basically acts as an "initiation."[19]

Nevertheless, Kuhn builds his argument for the plurality

of science on the temporary nature of paradigms, and the fact that different branches of science use different paradigms. Kuhn wants to arrive at foundational concepts underpinning the history of all science. But he maintains that scientists have the power to choose which paradigms they use and when.[20] In this sense, the success of a paradigm "can never be unequivocally settled by logic and experiment alone." This confirms that scientists' views and preferences play their part in the course of science.[21] Kuhn implicitly rejects grand concepts of unified science and theories of everything, wondering "whether truth in the sciences can ever be one."[22] Scientific revolutions occur within specialized traditions and do not need to apply to others.[23]

In the seventeenth century, scientists made many of the discoveries about gravity and the laws of motion we now take for granted. That was undoubtedly a grand scientific revolution. But Kuhn believes that scientific revolutions do not need to be so sweeping. For him, they come in different shapes and sizes, including both well-known "revolutions" such as that of the French chemist Antoine Lavoisier,* who discovered the existence of oxygen, or the less well-known discovery of the X-ray.[24]

All of these revolutions have one thing in common in Kuhn's view: they all break down and replace an existing paradigm.[25]

Overlooked

Scholars have often overlooked the psychological and cognitive points Kuhn makes in *Structure*.[26]

His social analyses of community affairs overshadow his

discussion of the "changes in perception" in Gestalt* psychology. The British historian of science Alexander Bird* notes that Kuhn knew more about psychology than sociology:[27] "[The] time has come to reappraise those naturalistic elements of Kuhn's thought that he himself abandoned such as the psychological nature of a scientific revolution and a psychological rather than linguistic notion of incommensurability."[28] Bird specifically referred to cognitive science* as providing tools for such analyses. But he did not elaborate on this idea.

Scholars working on the (overlooked) concepts of cognitive science in Kuhn's work include the philosophers of science Hanne Andersen,* Peter Barker,* and Xiang Chen*[29]; all have focused on Kuhn's less-discussed endorsement of the philosopher Ludwig Wittgenstein's* theory of "family resemblance." Wittgenstein argued that objects seemingly connected by one common feature may actually have a series of overlapping features tying them together[30]—that idea lay behind Kuhn's notion of a paradigm. According to Andersen, Barker and Chen, Kuhn was the only philosopher of science in the English-speaking world to adopt and interact with this idea before the mid-1970s.[31]

---

1. Thomas S. Kuhn, *The Structure of Scientific Revolutions*, 4th ed. (Chicago, IL: University of Chicago Press, 2012), x.
2. Kuhn, *Structure*, 206.
3. Kuhn, *Structure*, 20–1.
4. Kuhn, *Structure*, 37.

5. Kuhn, *Structure*, 163.
6. Kuhn, *Structure*, 38.
7. Kuhn, *Structure*, 40.
8. Kuhn, *Structure*, 79.
9. Alexander Bird, *Thomas Kuhn* (Chesham: Acumen, 2000), 3–9.
10. Kuhn, *Structure*, 68.
11. Kuhn, *Structure*, 69.
12. Kuhn, *Structure*, 169.
13. Kuhn, *Structure*, 163.
14. Kuhn, *Structure*, 168.
15. Kuhn, *Structure*, 169.
16. Kuhn, *Structure*, 171.
17. Kuhn, *Structure*, 95.
18. Kuhn, *Structure*, 166.
19. Kuhn, *Structure*, 164.
20. Kuhn, *Structure*, 50.
21. Kuhn, *Structure*, 95.
22. Kuhn, *Structure*, 167.
23. Kuhn, *Structure*, 50–1.
24. Kuhn, *Structure*, 93.
25. Kuhn, *Structure*, 92.
26. Alexander Bird, "*The Structure of Scientific Revolutions*: An Essay Review of the Fiftieth Anniversary Edition," *British Journal for the Philosophy of Science* 63, no. 4 (2012): 865.
27. Bird, "*The Structure of Scientific Revolutions*," 7–8.
28. Bird, *Thomas Kuhn*, 14.
29. Hanne Andersen, Peter Barker and Xiang Chen, *The Cognitive Structure of Scientific Revolutions* (Cambridge: Cambridge University Press, 2006), 12–8.
30. Nancy Nersessian, "Kuhn, Conceptual Change and Cognitive Science," in *Thomas Kuhn*, ed. Thomas Nickles (Cambridge: Cambridge University Press, 2003), 180.
31. Andersen, Barker and Chen, *The Cognitive Structure*, 8; Nersessian, "Kuhn and Cognitive Science," 1.

MODULE 7
ACHIEVEMENT

KEY POINTS

- *The Structure of Scientific Revolutions* achieved Thomas Kuhn's aim of historically re-examining the natural sciences.
- The rise in externalist* readings of science and of sociology* (according to which politics, economics and culture play a significant role in scientific advancement) did most to assist Kuhn's widespread success.
- Scholarly misunderstandings of Kuhn's work hampered his ability to convey his argument to a wide audience.

Assessing the Argument

Thomas Kuhn's *The Structure of Scientific Revolutions* chiefly aimed to challenge the view of science as a rational* exercise that continuously increases the quantity of human knowledge.[1]

Although it has been argued that Kuhn also wanted to bring about the demise of the philosophy of logical empiricism,* the science historians Michael Friedman* and George A. Reisch* have shown that he had strong ties with the empiricists.[2] Indeed, Kuhn's work shows astonishing similarities with that of the German logical-empiricist Rudolf Carnap.*[3] So it is accurate to say that while Kuhn's essentially historical view challenged the logical-empiricist mainstream of scholarship on the history and philosophy of science,* he accepted some of the central theoretical features of this "textbook tradition."

Seeing himself as the voice of the latest winning scientific faction,⁴ Kuhn hoped his alternative point-of-view would free the history of science* from the assumption that its only role was to reflect philosophy.

Uncovering Kuhn's true aims in his book is a much more difficult task than it appears at first glance; although he clearly states that he wants to attack the "textbook tradition" his intentions beyond that are not always apparent. This ambiguity has led critics to various interpretations. In the revolutionary era of the 1960s, many were willing to understand Kuhn as a radical and subversive thinker. Although he disavowed such portrayals, many passages appear to encourage such a reading. When the book was first published it made a revolutionary contrast to the dogmatic rigidity of mainstream scientific textbooks.⁵

We can see today that Kuhn's work produced profound changes in our understanding of science. But it is difficult to disentangle our knowledge of those changes from what Kuhn intended when he was writing the book. Although Kuhn challenged the common assumptions regarding science in his period, he may not have intended to challenge science itself—which is what eventually happened.⁶

In challenging the older, cumulative understanding of science, Kuhn certainly demonstrated that science is cyclical. But it remains difficult to claim that he accomplished anything more. Part of this difficulty stems from the vagueness of some of Kuhn's key concepts ("paradigm,"* for example). In addition, his work became so quickly and thoroughly adopted by social scientists and historians of science that it makes it difficult to imagine any pre-

Kuhnian model for understanding the history and development of science.

> "Kuhn's book was clearly and powerfully written, filled with persuasive examples, devoid of arcane vocabulary and symbols which had made philosophy of science a closed book to most laymen. Kuhn's ideas were quickly taken up by scholars in fields far removed from the natural sciences."
> ——Peter Novick, *That Noble Dream*

Achievement in Context

In the mid-twentieth century, universities increased their exploration of "externalist" readings of science and its place in society by paying attention to the greater social context in which scientific research occurs. The historian Richard Hofstadter's* work *Social Darwinism in American Thought 1860–1915* (1944) is a classic externalist examination. Hofstadter explored how the phrases "survival of the fittest" and "the struggle for existence" had emerged from the evolutionary biology of Charles Darwin,* entered into social commentary, and fed back into scientific practice in unexpected ways.[7] This trend certainly contributed to the success of *The Structure of Scientific Revolutions*, feeding the widespread interest that greeted the book when it was published.

In this regard, mention must be made of sociology—a field particularly concerned with the ways in which science and ideas have been historically shaped by social conditions. The historian of science James Marcum,* for example, believes that no discipline did more than sociology to support Kuhn's work, not least because of the way Kuhn examined the scientific community* as a social group,[8] looking within that community for the causes and characteristics of science.

In addition, *The Structure of Scientific Revolutions* appeared just as public debate about the use of atomic bombs in World War II became more widespread. Kuhn, who had been affected by this issue during his youth, wrote *Structure* during the height of the Cold War* at a time when anyone who critiqued science from a social or externalist viewpoint; many professors holding (or suspected of holding) such views lost their jobs.[9]

Kuhn represented a generation of new academics willing to shake up the scientific status quo and question the mainstream view of science as objective and certain. The time was ripe for this debate. As the British science interpreter Ziauddin Sardar* and the Canadian philosopher of science Ian Hacking* have pointed out, during the 1960s the public and the intellectual community began to question the role scientists, such as Kuhn's early mentor James Bryant Conant,* had played in creating atomic weapons.[10] In this sense, Kuhn's work marked a profound shift in the way we view science in society.

Structure has served as a springboard for new ideas about science. It marks an important milestone in the development of new scholarly and intellectual movements that have made science a function of scientists' social and intellectual backgrounds.[11] These thinkers questioned modern science in entirely new ways, emphasizing it as a product of a specific geographical area

(Western Europe), a specific civilization (Judeo-Christian), and a specific gender (male). Some even questioned science's claim to truth and knowledge.

Many of these ideas surprised Kuhn and some caused him concern. He had never imagined that his work had the potential to spark a wide-ranging re-evaluation of science and non-science subjects, such as the sociology of knowledge. The set of analytical tools his work provided found far wider applications than he could have anticipated.

As Sardar points out, *Structure* was primarily interested in the internal reform of science. Kuhn upheld science's autonomy from society. In his subsequent interventions to "correct" the misconceptions generated by his book, Kuhn aimed to insulate science from public scrutiny.[12] In fact, the American philosopher Steve Fuller* accused *Structure* of promoting conservative forms of scholarship and perpetuating preconceptions of the special nature of science. Fuller argued that some have used the text to marginalize radical philosophies of science. Two examples of this are the philosophies promoted by the scholars Jerry Ravetz* and Paul Feyerabend,* both of whom argued that science results entirely from the predilections and preferences of individual scientists acting without reference to or desire to form part of a community.*[13]

Limitations

Despite *Structure*'s continued and extraordinary popularity, it is worthwhile considering its limitations. Foremost among these is the work's defense of the special status bestowed on science by successive governments and societies in the twentieth century—a defense based on assumptions Kuhn inherited from the school in the philosophy of science known as logical empiricism.*

If Kuhn is right, however, and scientific knowledge is not wholly objective, being a product of the community that produced it, how then can it deserve or retain that special status? Stripped of it, science would be subject to the financial pressures of restricted university funding and the lure of financial rewards. How then will scientific research maintain its rigor? Pharmaceutical corporations offer high rewards for scientists who can come up with new drug compounds. Does this skew the research agenda in the scientific community? Kuhn did not consider these implications.

In addition, Kuhn offers only a preliminary exploration of the ways in which the scientific community shapes scientific practice.There is no discussion of the possible fragility of the social structures inside the scientific community in the face of a turbulent modern economy—as was exposed in the oil crisis of 1973,* when mounting oil prices brought economic instability. For example, universities faced tough choices over increasingly scarce resources, which affected the social aspects of the academic environment. But although the internal culture of the scientific community is not as stable as Kuhn seems to assume, differing over time and across territory,[14] readers will not find this factored into Kuhn's account. As a result, the historical analysis on which rest his concepts of paradigm and incommensurability* remains incomplete, as the special status science once enjoyed recedes.[15]

1. Thomas S. Kuhn, *The Structure of Scientific Revolutions*, 4th ed. (Chicago, IL: University of Chicago Press, 2012), x.
2. Michael Friedman, "Kuhn and Logical Empiricism," in *Thomas Kuhn*, ed. Thomas Nickles (Cambridge: Cambridge University Press, 2003), 19–21; George A. Reisch, "Did Kuhn Kill Logical Empiricism?," *Philosophy of Science* 58 (1991): 266–7.
3. Gurol Irzik and Teo Grunberg, "Carnap and Kuhn: Arch Enemies or Close Allies?," *British Journal for the Philosophy of Science* 46, no. 3 (1995): 305.
4. Kuhn, *Structure*, 135–8.
5. Kuhn, *Structure*, 166.
6. Peter Godfrey-Smith, *An Introduction to the Philosophy of Science: Theory and Reality* (Chicago, IL: University of Chicago Press, 2003), 99.
7. Richard Hoftstadter, *Social Darwinism in American Thought, 1860–1915* (Boston, MA: Beacon, 1944), 6.
8. James A. Marcum, *Thomas Kuhn's Revolution: An Historical Philosophy of Science* (London: Continuum, 2005), 142.
9. Ziauddin Sardar, "Thomas Kuhn and the Science Wars," in *Postmodernism and Big Science*, ed. Richard Appignanesi (Cambridge: Icon Books, 2002), 197.
10. Sardar, "Thomas Kuhn and the Science Wars," 195; Hacking, "Introductory Essay," in Kuhn, *Structure*, ix.
11. Sardar, "Thomas Kuhn and the Science Wars," in *Postmodernism and Big Science*, 211–21.
12. Sardar, "Thomas Kuhn and the Science Wars," 221–4.
13. Steve Fuller, *Thomas Kuhn: A Philosophical History for Our Times* (Chicago, IL: University of Chicago Press, 2000), 212.
14. John M. Ziman, *Real Science: What It Is and What it Means* (Cambridge: Cambridge University Press, 2000), 4–5.
15. Ziman, *Real Science*, 1.

MODULE 8
PLACE IN THE AUTHOR'S WORK

KEY POINTS

* Thomas Kuhn's work focused on the historical interpretation of scientific practice, with significant sociological* and philosophical speculation about underlying principles of scientific practice.
* *The Structure of Scientific Revolutions*, Kuhn's second book, set the trajectory for his later research.
* Kuhn's best-known, most widely debated and most-criticized work remains *The Structure of Scientific Revolutions*.

Positioning

It took Thomas Kuhn 15 years to write *The Structure of Scientific Revolutions*,[1] his second book. His first, *The Copernican Revolution* (1957), stemmed from his research into classical mechanics* (the physical laws concerning motion and forces as they were laid out before the twentieth century) and lacked the originality displayed in *Structure*.

In 1956, reviewing Kuhn's description of the yet-unpublished *The Copernican Revolution* on his fellowship application form, Harvard University's Tenure Committee deemed the work less a contribution to scientific knowledge than an attempt to popularize science.[2] He did not win tenure (the security of a permanent academic position). Containing early versions of ideas that would become central to *The Structure of Scientific Revolutions* and Kuhn's later work, the book nevertheless interests Kuhn scholars.

In the sixteenth century the Polish astronomer Nicolaus Copernicus* formulated the theory that the sun, rather than the earth, sat at the center of the universe. For centuries, people had assumed that the success of this worldview depended on Copernicus's unique insights in the field of astronomy. Kuhn argued, however, that the social and intellectual climate in which Copernicus wrote also played a central role.[3] In researching his book, Kuhn was able to make a case study of revolutionary thought—a mindset that would become central to the idea of "extraordinary science"* in *The Structure of Scientific Revolutions*.

A second major component of Kuhn's scheme of scientific development in *The Structure of Scientific Revolutions* concerns the idea of normal science.* Kuhn first expressed this idea in the article "The Function of Measurement in Modern Physical Science" (1961). Here he presented normal science as a process of applying the gains made in scientific revolutions* to the practice of everyday science, arguing that "the function of measurement and the source of its special appeal are derived largely from myth."[4]

Kuhn had also already introduced the notion of the paradigm* in "The Essential Tension," a paper presented in 1959 at the University of Utah Research Conference.[5]

> "One of the primary forms of what is sometimes referred to as 'historicism,' sometimes as 'relativism,' is Thomas S. Kuhn's ***The Structure of Scientific Revolutions***. Although Professor Kuhn has frequently insisted that most such interpretations of his views have distorted his meaning, it is not entirely clear that he has successfully answered those of his critics

> who have thus interpreted his work, nor has he clarified his position so that the matter is no longer open to debate."
> ——Maurice Mandelbaum, "A Note on Thomas S. Kuhn's *The Structure of Scientific Revolutions*"

Integration

Despite his extensive body of subsequent work, *The Structure of Scientific Revolutions*, particularly its first edition of 1962, defines Kuhn today, creating an enduring vision of its author as a radical. Intellectual and popular circles have continued to feed this myth. But although the imagery and rhetoric in the first edition of *Structure* do seem radical, recent scholarship has shown that Kuhn later modified, dismantled, and underplayed many claims and implications in the work.

Overall, and throughout his published works, Kuhn maintains the same intention: to re-assess in historical perspective and with reference to the scientific community* how science has changed and developed. His emphasis on the non-scientific aspects of science and on the historical dynamics of scientific development never changed. Nor did his commitment to the idea of paradigm and the thesis of incommensurability,* despite heavy and sustained criticism. But the seemingly radical language in the first edition of *The Structure of Scientific Revolutions* contrasts with Kuhn's later writings.[6]

Scholars still debate this point: did Kuhn retreat from a radical position after his clashes with philosophers of science? Or did the revolutionary tone of his rhetoric overstate his intentions?

Significance

The Structure of Scientific Revolutions made Kuhn famous. It quickly became an influential text not only for the history and philosophy of science* but for other disciplines as well. And although Kuhn spent much of his career after 1970 clarifying aspects that he felt critics had misunderstood,[7] he continued to believe that *The Structure of Scientific Revolutions* remained his best work.[8]

After *Structure*, Kuhn turned to research in pure history of science,* which he pursued into the late 1970s. *Black-Body Theory and the Quantum Discontinuity, 1894–1912* (1978) was a straightforward historical analysis and made no mention of the ideas of *Structure*.[9] He discussed some of the work of the early twentieth-century German theoretical physicist* Max Planck.* Critics had long viewed Planck's 1900 and 1901 papers on quantum physics* as having catalyzed the transition from classical to quantum mechanics.* Kuhn took the opposite view,[10] however, arguing that Planck had in fact never shed his classical mechanics* worldview.

In hindsight, Kuhn believed *Black-Body Theory* to have been his best work on a specifically historical subject. Historians, philosophers, and physicists, however, did not agree, some bemoaning the fact that Kuhn did not use the framework he laid out in *The Structure of Scientific Revolutions*.[11] In response to the criticism, Kuhn added an afterword in a revised edition that sought to clarify the relationship of this work to the ideas in *Structure*.[12]

After 1978, Kuhn returned once more to the concept of

incommensurability, this time looking at it through the framework of linguistics. In doing so, he sought to distance himself from Gestalt psychology* and its theories of perception. According to Gestalt psychology, individuals construct their perceptions—how they see, respond to, and interact with the world—around a theoretical framework based on their beliefs. Kuhn thought the connection people had made between his theories and Gestalt had led people to misunderstand his views of incommensurability. Too many scholars saw him as arguing that scientists simply construct competing paradigms without reference to reality. In this scenario, the incommensurability of paradigms could appear to derive solely from people's arguments.

Kuhn wanted to reverse scholars' misinterpretation of his work, so he set about lecturing on areas of *Structure* that had caused confusion. For example, although he never published a book on the topic, he lectured on the idea that links exist between scientific knowledge of reality and the structure of language.[13] His theory was that a scientific paradigm is linked to a vocabulary of concepts and terms specific to that paradigm. But this vocabulary is also specific to the field of science and the scientific community in which it operates. So scientists working in different fields, and using separate paradigms, cannot borrow from other paradigms to assist each other in their research or to collaborate because the scientific vocabulary and conceptual framework of all paradigms is unique. This is why there is so much plurality within "science"—in reality, singular science is a collection of sub-disciplines and communities of scientists each with their own paradigms.[14]

In 1977 Kuhn further expanded his ideas on the incommensurability thesis by discussing in detail the way scientists choose which paradigm to adopt as the consensus—a process he labeled "theory choice." He thought the concept was ripe for further explanation after becoming aware of *Structure*'s shortcomings. In particular, he had not addressed this question: "If scientists ultimately chose a consensus paradigm* for reasons other than purely scientific considerations, could their choice be irrational?"

Wanting to avoid the implication of irrationality,* because it clashed with his intention to preserve science as an autonomous, special activity,[15] Kuhn re-asserted and elaborated on five scientific considerations that guide those in the scientific community with the power and authority to decide on a consensus paradigm:

- Accuracy
- Consistency
- Breadth of scope
- Simplicity
- Fruitfulness[16]

So although sociological features of the scientific community decided which scientists had the power to choose the consensus paradigm, scientific considerations molded their final choice.

---

1. Thomas S. Kuhn, *The Structure of Scientific Revolutions*, 4th ed. (Chicago, IL: University of Chicago Press, 2012), xxxix.
2. James A. Marcum, *Thomas Kuhn's Revolution: An Historical Philosophy of Science* (London: Continuum, 2005), 13–4.

3. Thomas S. Kuhn, *The Copernican Revolution: Planetary Astronomy in the Development of Western Thought* (Cambridge, MA: Harvard University Press, 1957), viii.
4. Thomas S. Kuhn, "The Function of Measurement in Modern Physical Science," in *The Essential Tension: Selected Studies in Scientific Tradition and Change* (Chicago, IL: Chicago UP, 1979), 161.
5. Thomas S. Kuhn, "The Essential Tension" in *The Essential Tension*, 18–26.
6. Peter Novick, *The Noble Dream: The "Objectivity Question" and the American Historical Profession* (Cambridge: Cambridge University Press, 1988), 526–8.
7. Thomas S. Kuhn, *The Road Since Structure: Philosophical Essays, 1970–1993 with an Autobiographical Interview*, ed. James Conant and John Haugeland (Chicago, IL: Chicago University Press, 2000), 7.
8. Marcum, *Thomas Kuhn's Revolution*, 109.
9. Thomas S. Kuhn, *Black Body Theory and Quantum Discontinuity, 1894–1912* (Chicago, IL: Chicago UP, 1978).
10. See Marcum's short and useful summary of this work in *Thomas Kuhn's Revolution*, 108–12.
11. Marcum, *Thomas Kuhn's Revolution*, 109.
12. See Kuhn, *Black Body Theory* 2nd ed. (1987), 349–79.
13. Marcum, *Thomas Kuhn's Revolution*, 24.
14. Howard Sankey, "Kuhn's Changing Concept of Incommensurability," *British Journal for the Philosophy of Science* 44, no. 4 (1993): 770–2 describes the changes in Kuhn's theory of incommensurability.
15. Kuhn, "Objectivity, Value Judgement and Theory Choice," in *The Essential Tension*: 321–2.
16. Kuhn, "Objectivity, Value Judgement and Theory Choice," 330–9.

SECTION 3
IMPACT

MODULE 9
THE FIRST RESPONSES

KEY POINTS

- Critics initially focused on Kuhn's concepts of paradigm* and incommensurability,* the relativism* they perceived in his work (that is, roughly, the notion that "perfect" knowledge is impossible to achieve), and the image of the scientific community* he presented.
- The most influential criticism came from the philosophers Dudley Shapere* and Karl Popper,* who seized on the lack of precision in Kuhn's concept of paradigm. Popper in particular attacked Kuhn's understanding of scientific progress as cyclical.
- Many of the earliest responses were shaped by the cultural and political status of science as part of a free society in the Cold War.*

Criticism

In the first two years after its appearance, Thomas Kuhn's *The Structure of Scientific Revolutions* received positive reviews. After 1964, critical voices began to appear, chiefly from philosophers of science. Critics concentrated on three aspects, focusing on the term "paradigm" (and, by association, the way *Structure* portrayed science and the scientific community), what they perceived as the text's relativism, and idealism.*

In 1964, the philosopher of science Dudley Shapere criticized Kuhn for his definition of paradigm, deeming the notion too imprecise.[1] He also had trouble with Kuhn's suggestion that a paradigm shift* could alter the actual meaning of a scientific concept such as "mass." He thought instead it might alter how that notion would be applied.[2] Finally, Shapere found the ideas of incommensurability and paradigm impossible to square. Understanding incommensurability to mean complete "world change," Shapere noted that paradigms could not really disagree among themselves.[3]

The influential Austrian British philosopher Karl Popper also criticized Kuhn for the lack of precision in his term "paradigm." At a 1965 seminar at the London School of Economics, Popper and his followers clashed with Kuhn.[4] The pro-Popper philosophers of science, who included Kuhn's Berkeley colleague Paul Feyerabend,* pilloried Kuhn. But it is difficult to know how much this argument impacted on Kuhn's reputation at large. Since, as one commentator observed, in 1965 "both Kuhn's and Popper's views on *science* were probably known more by reputation than readership,"[5] the discussion likely never included more than a small group of scholars. In Popper's view, science by its nature had revolutionary qualities. Its fundamental characteristics included "bold conjectures" and constant retesting and refutations of its theories.[6] Popper and others at the conference disagreed profoundly with Kuhn's division of science into "normal" and "extraordinary." For them, as Kuhn's interpreter and defender Paul Hoyningen-Huene* points out, science is an enterprise "molded by the persistent awareness of the fallibility of human epistemic claims."[7]

Popper went on to decry Kuhn's idea of normal science* as a "danger to science and indeed to our civilization."[8] In Popper's view, the dogmatic character of Kuhn's ideas was profoundly unscientific.⁹ Although Popper acknowledged that dogmatic enterprise existed, he saw "normal science" as a contradiction in terms.¹⁰ Followers of Popper also remained dissatisfied with Kuhn's depiction of the scientific community. Believing science to be fundamentally open, they harshly criticized Kuhn for describing it as "a closed society whose chief characteristic is 'the abandonment of critical discourse.'"¹¹

The strongest criticism raised by philosophers of science was that Kuhn supported irrationality* and relativism. As the scholars Vasso Kindi* and Theodore Arabatzis,* interpreters and editors of Kuhn's work, have recently pointed out, these philosophers feared that Kuhn's argument was overly broad. Kuhn had said that considerations outside the scope of scientific knowledge (the professional rivalries and power struggles of the scientific community and the personal views of scientists, for example) determined the methods and conclusions of scientific knowledge. Many philosophers feared this went too far.¹² They disliked Kuhn's argument in principle because it compromised the claims of science to stand above scholarly disputes in a realm of unquestionable objectivity.

As Popper saw it, Kuhn was arguing that scientists could not rationally decide which framework to follow.¹³ The Hungarian philosopher of science Imre Lakatos* accused Kuhn of proposing a view of scientific change characterized by the views of "mob psychology." He also decried Kuhn's view of paradigm change as a "mystical conversion, which is not and cannot be governed by rules of reason."¹⁴

> "'Look,' Thomas Kuhn said. The word was weighted with weariness, as if Kuhn was resigned to the fact that I would misinterpret him, but he was still going to try—no doubt in vain—to make his point. 'Look,' he said again. He leaned his gangly frame and long face forward, and his big lower lip, which ordinarily curled up amiably at the corners, sagged. 'For Christ's sake, if I had my choice of having written the book or not having written it, I would choose to have written it. But there have certainly been aspects involving considerable upset about the response to it.'"
>
> ——John Horgan, *The End of Science*

Responses

Very sensitive to the criticism of his peers, Kuhn responded by delivering papers at academic conferences in London (1965), his Swarthmore Lecture (also in London, 1967) and at the Urbana conference in 1969.[15] He maintained that he had not intended to uphold relativism and irrationality. In a long response, he affirmed that science "is our surest example of sound knowledge." He also expressed his belief in the progress of science, albeit not according to the traditional perspective.[16]

As in *The Structure of Scientific Revolutions*, he sought to clarify this idea by making analogies with evolution. Charles Darwin* theorized that the fitness of a species determines its continued survival. Just as in Darwin's evolution, scientific evolution moves with no clear goal, irreversibly, in only one direction. Kuhn particularly rejected the idea that scientific development brings knowledge closer to the truth.[17]

Additionally, Kuhn strongly defended the concept of normal science in both a descriptive and normative* sense (that is, telling the reader how science ought to be rather than how it actually was). He claimed that we can detect revolutionary science* only against the background of normal science. He also maintained that the static state of normal science served to deepen scientific research and progress. In a provocative essay published in 1963, before his 1965 confrontation with Karl Popper and his supporters, Kuhn had defended the function and importance of "dogma" for scientific research. He later abandoned the use of this word due to its negative connotations.[18]

Kuhn also made great efforts to clarify the notion of the paradigm—a concept that critics had viewed as everything from unclear to downright incoherent. He admitted that his initial use of the term had been too vague, and that the "excessive plasticity" of the term "paradigm" in *The Structure of Scientific Revolutions* had led to inappropriate applications of the notion to other disciplines, particularly sociology.[19] He sought to clarify the term by dividing it into a "broad" and a "narrow" sense, likening the broad sense to a disciplinary matrix* (the symbolic generalizations, models, values, and problem solutions employed by a scientific community[20]) and the narrow sense to the solutions to concrete problems.

The German philosopher of science Paul Hoyningen-Huene observed that Kuhn stopped using the term disciplinary matrix in his later work, as he focused on the notion of the exemplar*—the idea that during periods of normal science, scientists' solutions to particular problems become well-known as models ("exemplars") of conduct under the consensus paradigm.*[21] Another change involved Kuhn's realization that "paradigm" could refer both to universal consensus and the internal consensus of certain schools of thought.

Kuhn also tried to clarify the concept of incommensurability, wanting to discourage radical understandings of the notion. In the initial edition of *Structure*, Kuhn offered a somewhat ambiguous explanation of incommensurability as "world changes." This caused readers to assume that Kuhn intended a paradigm shift to imply the complete displacement of a conceptual framework.[22] In fact, Kuhn later claimed that he never meant an entire conceptual framework change—only some theories, terms, vocabularies, or languages.[23] Communication remains possible, then, but it is only partial.

This approach led Kuhn to further emphasize language and translation problems.[24] At this point he referred to the work of the philosopher of science V. O. Quine,* whose theory of indeterminacy asserted that translations could never be perfect.[25] Kuhn used Quine's argument to show that translation of all terms across paradigms is incomplete.At the same time, some terms remained the same, and this was where communication could continue.

Kuhn intended these modifications to clarify that the choice of a new theory is not fundamentally irrational; some aspects, he proposed, remained apt for comparison. Among these were empirical* predictions and assertions about specific situations (that is, predictions and assertions based on observable evidence and rational deductions), and the comparison of two competing theories.[26] He strongly denied that incommensurability and incomparability were the same thing.[27]

Conflict and Consensus

Although the modifications Kuhn made to *The Structure of Scientific Revolutions* dramatically transformed his work, they did not appease all of his critics, many of whom saw Kuhn's clarifications as a new and less radical theory.[28]

The philosopher of science John Worrall,* who had studied under Imre Lakatos, criticized Kuhn's concession on the subject of progress. Worrall argued that watering down the concept of incommensurability detracted from his overall argument.[29] Kuhn remained adamant, however, that he intended his explanations to clarify—not modify—his ideas.

Some critics were not completely appeased by the revised edition of *Structure* that Kuhn published in 1970. The philosophers of science Alan Musgrave* and Dudley Shapere, for example, remained unconvinced by Kuhn's use of a disciplinary matrix and exemplar to explain paradigms.[30] Shapere, arguably Kuhn's strongest critic, felt that Kuhn had not successfully defended himself against the charge of relativism.[31] The underlying problem derived from the first edition of *Structure*, in which Kuhn's key concepts such as paradigm, so Shapere believed, remained obscure. And one cannot produce an intelligible concept by further explaining the nonsensical.[32]

The controversy surrounding *The Structure of Scientific*

Revolutions did not die down quickly. Although it became less heated in the following years, Kuhn continued to be the subject of criticism, especially from philosophers of science. Kuhn even angered his own supporters, particularly radical scholars and thinkers, displeased by his later retractions.

Nevertheless, he remained adamant that writing *Structure* proved a timely contribution to the history and philosophy of science,* commenting when interviewed that "if I had my choice of having written the book or not having written it, I would choose to have written it. But there have certainly been aspects involving considerable upset about the response to it."[33]

---

1. Dudley Shapere, "The Structure of Scientific Revolutions," *Philosophical Review* 73, no. 3 (1964): 388.
2. Shapere, "The Structure," 390.
3. Shapere, "The Structure," 391.
4. The story of the confrontation between Kuhn and Popper is well captured in Steve Fuller, *Kuhn vs Popper: The Struggle for the Soul of Science* (New York: Columbia University Press, 2004).
5. Fuller, *Kuhn vs Popper*, 29.
6. Karl Popper, "Normal Science and Its Dangers," in *Criticism and the Growth of Knowledge*, ed. Imre Lakatos and A. Musgrave (Cambridge: Cambridge University Press, 1970), 55.
7. Paul Hoyningen-Huene, *Reconstructing Scientific Revolutions: Thomas S. Kuhn's Philosophy of Science* (Chicago, IL: University of Chicago Press, 1993), 168.
8. Popper, "Normal Science," 53.
9. James A. Marcum, Thomas Kuhn's *Revolution: An Historical Philosophy of Science* (London: Continuum, 2005), 86–7.
10. John Worrall, "Normal Science and Dogmatism, Paradigms and Progress: Kuhn 'Versus' Popper and Lakatos," in *Thomas Kuhn*, ed. Thomas Nickles (Cambridge: Cambridge University Press, 2003), 67–9.
11. J. Watkins, "'Against "Normal Science"'", in *Criticism and the Growth of Knowledge*, ed. Lakatos

and Musgrave, 37.
12. Vasso Kindi and Theodore Arabatzis, "Introduction," in *Kuhn's The Structure of Scientific Revolutions Revisited*, ed. Vasso Kindi and Theodore Arabatzis (New York: Routledge, 2012), 2.
13. Marcum, *Thomas Kuhn's Revolution*, 87.
14. Imre Lakatos, "Falsification and the Methodology of Scientific Research Programmes," in *Criticism and the Growth of Knowledge*, ed. Lakatos and Musgrave, 93.
15. Marcum, *Thomas Kuhn's Revolution*, 101.
16. Thomas S. Kuhn, "Logic of Discovery or Psychology of Research?," in *Criticism and the Growth of Knowledge* ed. Lakatos and Musgrave, 20.
17. Kuhn, "Logic of Discovery," in *Criticism and the Growth of Knowledge*, ed. Lakatos and Musgrave 1.
18. Thomas S. Kuhn, "The Function of Dogma in Scientific Research," in *Scientific Change*, ed. Alistair C. Crombie (New York: Basic Books, 1963), 347–69.
19. Thomas Kuhn, "Second Thoughts on Paradigms," in *The Essential Tension: Selected Studies in Scientific Tradition and Change* (Chicago, IL: University of Chicago Press, 1977), 259, 295–319.
20. See Paul Hoyningen-Huene's detailed analysis of the components of the disciplinary matrix in *Reconstructing Scientific Revolutions*, 145–59.
21. Hoyningen-Huene, *Reconstructing Scientific Revolutions*, 143.
22. W.H. Newton-Smith, *The Rationality of Science* (London: Routledge, 1981), 12; Hilary Putnam, *Reason, Truth and History* (Cambridge: Cambridge University Press, 1981), 115.
23. Thomas S. Kuhn, "Postscript," *The Structure of Scientific Revolutions*, 4th ed. (Chicago, IL: University of Chicago Press, 2012), 197–203.
24. Howard Sankey, "Kuhn's Changing Concept of Incommensurability," *British Journal for the Philosophy of Science* 44, no. 4 (1993): 765.
25. Kuhn, "Postscript," *Structure*, 201–2.
26. Hoyningen-Huene, *Reconstructing Scientific Revolutions*, 219–21.
27. Thomas S. Kuhn, "Theory-Change as Structure-Change: Comments on the Sneed Formalism," *Erkenntnis* 10, no. 2 (1976): 191.
28. Newton-Smith, *The Rationality of Science*, 113–4; Putnam, *Reason, Truth and History*, 126; M.V. Curd, "Kuhn, Scientific Revolutions and the Copernican Revolution," *Nature and System* 6 (1984): 4.
29. Worrall, "Normal Science," 93.
30. Musgrave, "Kuhn's Second Thoughts," 293; Shapere, "The Paradigm Concept," *Science* 172 (1971): 707.
31. Shapere, "The Paradigm Concept," 708.
32. Shapere, "The Paradigm Concept," 710.
33. Peter Godfrey-Smith, *An Introduction to the Philosophy of Science: Theory and Reality* (Chicago, IL: University of Chicago Press, 2003), 87.

MODULE 10
THE EVOLVING DEBATE

KEY POINTS

* *The Structure of Scientific Revolutions* became a point of reference for sociologists of science and working scientists themselves as an ever-broadening array of scholars debated scientific concepts, often out of their context.
* Although no Kuhnian school of thought exists today, his work has had wide-ranging impact of history work in philosophy, history, sociology,* psychology, and in the natural sciences.
* Kuhn's work has given rise to a new debate within the sub-discipline of the sociology of scientific knowledge,* exploring how scientists' tastes affect their conclusions.

Uses and Problems

Thomas Kuhn's *The Structure of Scientific Revolutions* is regarded by scholars as a seminal work both for those studying the history of science,* for students of the sociology and philosophy of knowledge (and sociology more generally), and for those working in the educational and non-profit sectors to promote public engagement with science.

Like his contemporaries, the philosophers Paul Feyerabend,* Karl Popper* and Norwood Russell Hanson,* Kuhn presented radical reinterpretations of science that ended the domination of the logical-empiricist* school of thought. More than any of his contemporaries, "Kuhn changed the philosophy of science* by describing a tremendously vivid picture of scientific change."[1]

Although *Structure* generated considerable controversy when it was first published, it is difficult, perhaps, to detect the impact of the ideas and problems it offered in the work of subsequent scholars.

The philosopher of science Ian Hacking* argued in 1981 that *The Structure of Scientific Revolutions* put an end to several dominant concepts, the most notable being, arguably, that of realism*—the idea that science can discover truths about the real world.[2] Other scholars find that *Structure* has had a minimal impact on the philosophy of science. Hanne Andersen,* Peter Barker,* and Xiang Chen,* for example, philosophers of science from Denmark, Britain, and the US respectively, argue that philosophers of science have developed negative perceptions of Kuhn's work because of mistakes that have persisted for nearly a half century. They believe that contemporary philosophy of science has withdrawn from historical studies; instead, it concentrates on defending the realist position (that scientific facts can only be determined by observation) against the constructivist* position (that scientists determine facts by observing reality and interpreting it in accordance with their beliefs.)[3]

> "The basic problem is that there are, as philosopher of science Tim Maudlin has eloquently pointed out, **two** Kuhns—a moderate Kuhn and his immoderate brother—jostling elbows throughout the pages of *The Structure of Scientific Revolutions."*
> ——Alan Sokal and Jean Bricmont, *Intellectual Impostures*

Schools of Thought

Kuhn's work falls between currents of thought in the history and philosophy of science and no strictly Kuhnian school of thought exists. Scholars agree that Kuhn's work proved highly provocative for historians of science, even if those historians adopted few of its propositions. The American historian of philosophy Jan Golinski,* for example, claims that Kuhn's "influence among historians has been at best limited"[4] (a view shared by the scholars Vasso Kindi* and Theodore Arabatzis*).[5]

These views have persisted despite the efforts of Kuhn's own students to spread his approach to the history of science. The American academic John L. Heilbron,* for example, wrote that Kuhn "gave us to understand that we were engaged in an intellectual adventure of great moment."[6] Historians of science—especially N. M. Swerdlow,* J. Z. Buchwald* and Norton Wise*—took up Kuhn's notions of the scientific community,* scientific revolutions* and incommensurability* to further his research agenda and the cyclical understanding of scientific progress.[7]

In the 1970s, the movement away from Kuhn's work gave rise to externalist* approaches to the way scientists conduct research. These approaches emphasized that science is the product of external factors such as society and political events, and paid less, if any, attention to the requirements of scientific accuracy and integrity.

In the 1990s, the field moved on to new concerns, for which Kuhn had no answers. Kuhn's critique of the older understandings of science ironically led to the discipline dismissing him as representative of the "outmoded genre of grand narratives."[8] In other words, historians chose to abandon internalist* assumptions such as those Kuhn highlighted when explaining how scientists chose the consensus paradigm,* turning instead to new methods, particularly from sociology, seeking to convey more than the "pallid* platitudes" about science seen from the inside.[9]

This movement has continued into the 1990s and early 2000s with the rising tide of externalist readings focused on the work of the influential French philosopher Michel Foucault,* who saw social determinants as paramount in the search for understanding.[10]

In Current Scholarship

Currently, radical intellectuals remain the most avid proponents of Thomas Kuhn's text. *The Structure of Scientific Revolutions* has become a foundational piece of writing for postmodernist* critiques of science, which question the possibility of objectivity in science, considering it a form of cultural production like any other and subject to the same issues of context and interpretation. These approaches have spurred new sub-genres of scholarship such as feminist* and post-colonial* studies. These new disciplines seek to demonstrate that, although ignored by previous scholars, women and colonial subjects had a place in the history of science.[11] As the science interpreter Ziauddin Sardar* points out, Kuhn has been considered to be "subversive of science" by those eager to preserve what little remains of the internalist reading of science presented by logical empiricists (meaning those trying to save scientists' claims to objectivity while simultaneously exploring how personal and professional factors shape scientific practice).[12]

Kuhn's ideas remain central to the sociology of scientific knowledge,[13] an approach formulated by the sociologists David Bloor* and Barry Barnes* of the University of Edinburgh which maintains that social conditions create scientific knowledge.[14] The French sociologist Bruno Latour* and his colleague Stephen Woolgar* produced the most infamous example of this approach. "Our most general objective," they wrote in the preface to *Laboratory Life* (1969), "is to shed light on the nature of 'the soft underbelly of science': we therefore focus on the work done by a scientist located firmly at his laboratory bench."[15] By "soft underbelly" they mean the malleable social conditions that, they allege, shape science.

Most recently, postmodernist critics of science have turned their focus to the role of science in society. As Sardar has observed, science is a field that exercises considerable power and authority, remaining deeply tied up with globalized* corporate and government capital. These newest critics center their debate on the importance and level of public scrutiny of science.

Radical critics of science argue that today science has become so closely linked with political and economic power structures— government and global corporations—that historians and philosophers of science, as neutral or disinterested observers, need to scrutinize scientists' activity in the same way that politicians' and lobbyists' motives and conflicts of interest are examined. Such critics have also questioned the billions Western countries spend on scientific research each year.[16]

Some still support a vision of science as independent and autonomous. These critics counter that the "academic Left" cannot properly understand what science is and how it works. Embedded in this is a criticism of the language of multiculturalism.* The mathematician and physicist Alan Sokal* has said that, in an environment that prizes multiculturalism, "incomprehensibility becomes a virtue; allusions, metaphors and puns substitute for evidence and logic."[17] He argues that postmodernist scholarship has resulted in the loss of public faith in science—a loss, as he sees it, hindering progress. As the biologist E. O. Wilson* succinctly (and sarcastically) put it in a 1994 talk, "multiculturalism equals relativism* equals no supercollider."*[18] In other words, Sokal and Wilson argue that once the credibility of scientific work comes into question, as it has since Kuhn and especially with the work of postmodernists, then it loses funding from government, corporations, and academic foundations. This means, ultimately, that scientific experiment will not continue and so the contribution of scientific work to public health and living standards will collapse, leaving the world in a state of ignorance. This, they contend, represents a step backwards that should not take place—no matter how imperfect the current state of science may be.

---

1. Peter Godfrey-Smith, *An Introduction to the Philosophy of Science: Theory and Reality* (Chicago, IL: University of Chicago Press, 2003), 98.
2. Ian Hacking, *Scientific Revolutions* (Oxford: Oxford University Press, 1981), 1–2.
3. Hanne Andersen, Peter Barker and Xiang Chen, *The Cognitive Structure of Scientific Revolutions*

(Cambridge: Cambridge University Press, 2006), 238.
4. Jan Golinski, *Making Natural Knowledge: Constructivism and the History of Science* (Cambridge: Cambridge University Press, 1998), 14.
5. Vasso Kindi and Theodore Arabatzis, "Introduction," in *Kuhn's The Structure of Scientific Revolutions Revisited*, ed. Vasso Kindi and Theodore Arabatzis (New York: Routledge, 2012), 2.
6. John L. Heilbron, "A Mathematicians' Mutiny, with Morals," in *World Changes: Thomas Kuhn and the Nature of Science*, ed. Paul Horwich (Cambridge, MA: MIT Press, 1993), 112.
7. James A. Marcum, *Thomas Kuhn's Revolution: An Historical Philosophy of Science* (London: Continuum, 2005),134–6.
8. Kindi and Arabatzis, "Introduction," 2.
9. Kindi and Arabatzis, "Introduction," 3.
10. Peter Novick, *The Noble Dream: The "Objectivity Question" and the American Historical Profession* (Cambridge: Cambridge University Press, 1988), 536–7.
11. Ziauddin Sardar, "Thomas Kuhn and the Science Wars," in *Postmodernism and Big Science*, ed. Richard Appignanesi (Cambridge: Icon Books, 2002), 216–21.
12. Sardar, "Thomas Kuhn and the Science Wars," 221.
13. Barry Barnes, *T.S. Kuhn and Social Science* (London: Macmillan, 1982).
14. David Bloor, *Knowledge and Social Imagery*, 2nd ed. (Chicago, IL: Chicago University Press, 1991), 7,166.
15. Bruno Latour and Steve Woolgar, *Laboratory Life: The Construction of Scientific Facts* (New York: Sage Publications, 1979), 27.
16. Sardar, "Thomas Kuhn and the Science Wars," 6.
17. Alan Sokal, "A Physicist Experiments with Cultural Studies," *Lingua Franca* (1996), 62–4.
18. Quoted in Michael Bérubé, "The Science Wars Redux," *Democracy Journal* 19 (2011): 67.

MODULE 11
IMPACT AND INFLUENCE TODAY

KEY POINTS

- *The Structure of Scientific Revolutions* presents a contested understanding of scientific knowledge that continues to stimulate debate and misunderstanding across the social sciences and philosophy today.
- Scholars situate Kuhn's work between the logical positivist* school (with its emphasis on formal methods) and the relativist* school (with its assumption that ultimate, "perfect" solutions to scientific questions are impossible). But it poses challenges to each.
- Responses to Kuhn's work continued into the twenty-first century as the rise of postmodernism* prompted scholars to increasingly contest the notion that science is useful for other subjects.

Position

In 1993 the German American philosopher of science Carl Hempel* anointed Thomas Kuhn's *The Structure of Scientific Revolutions* a landmark history of science.* Hempel opened a collection of essays about the philosophy of science* by addressing Kuhn directly:"Whatever position your colleagues may take,Tom, I am sure that they all feel a large debt of gratitude to you for your provocative and illuminating ideas."[1] Hempel's tribute is especially noteworthy because Kuhn's work challenged the logical-empiricist* approach taken by Hempel.

Many scholars find that the work had the least impact where

Kuhn would have wanted it to have the most: on the history of science. In fact, recent critics allege "Kuhn's effect on science studies has been to dull the importance of history and paralyze the discussion of politics."[2]

Certainly the work, with its emphasis on history and culture, contributed to the demise of logical empiricism, a school founded on notions of objectivity and purely verifiable analysis. But its positive influence on the philosophy of science remains less clear. Although the unprecedented response to Kuhn's work has made it famous, it has also prevented *The Structure of Scientific Revolutions* from making a lasting and definable contribution to the subject Kuhn prized most: the history of science. In part, this must be because that response has focused more on the philosophical content of Kuhn's argument—concepts such as paradigm,* incommensurability* and exemplar*—rather than the historical reading of science on which Kuhn based these concepts.

> "Fifty years ago this month, one of the most influential books of the twentieth century was published by the University of Chicago Press. Many if not most lay people have probably never heard of its author, Thomas Kuhn, or of his book, ***The Structure of Scientific Revolutions***, but their thinking has almost certainly been influenced by his ideas.The litmus test is whether you've ever heard or used the term "paradigm shift," which is probably the most used—and abused—term in contemporary discussions of organisational change and intellectual progress. A Google search for it returns more than 10 million hits, for example. And it currently turns up inside no fewer than 18,300 of the books marketed by Amazon. It

> *is also one of the most cited academic books of all time. So if ever a big idea went viral, this is it."*
> —John Naughton, "Thomas Kuhn: The Man Who Changed the Way the World Looked at Science"

Interaction

The Structure of Scientific Revolutions remains part of the current intellectual debate on science. But Kuhn might not be comfortable with the direction of the debate today in which the battle lines are drawn between science realists* (who believe, roughly, that we arrive at scientific facts through observation) and constructivists* (who consider science a matter of interpretation in which the scientist's beliefs play a significant part).

The debate between these two factions exploded in the 1990s in the so-called "science wars."* In this period realist supporters—mainly practitioners and some philosophers of science—attacked as "irrational" cultural and postmodernist ideas, frequently held by people on the political Left. The American scientists Paul Gross* and Norman Levitt* fired the first shot in this "war" in 1994 when they published *Higher Superstition*, a book in which they decried postmodernism as "medieval" in outlook and condemned the academic Left's bias against science.[3] They followed this with a 1995 conference in New York City called "The Flight from Science and Reason." Papers presented at the conference condemned criticism of science as "common nonsense" and critics as "charlatans."[4]

In the 2000s, Gross, Levitt, and some of those they attacked on the Left took steps to end the argument. But the quarrel persists because it has a bearing on how universities distribute funds for scientific research—money lying at the root of this "war," as it does in so many real-life conflicts. To Gross and Levitt, challenging the concrete conclusions of science undermines its credibility and marginalizes it in universities.[5] On the Left, the American philosopher Michael Bérubé* argues that what postmodernists really object to is "fundamentalism" about science being untouchable. By drawing the battle lines against this view, all parties may be able to unite since Gross, Levitt, and Bérubé all desire the practice of high-quality science that is open to debate. All of them share with Kuhn a desire to improve the role of science in society through a better understanding of how it works; all differ from Kuhn in that, by opening science to debate, they take away its status as untouchable (something that Kuhn guarded).

Nevertheless, the issue of science's position in society does not necessarily translate into clear-cut political stances. As Bérubé noted: "The 1990s have not been kind to American institutions of higher education. Academy bashing is now amongst the fastest-growing of major US industries."[6] University funding in general came under attack in the twenty-first century, and this has added an additional layer of logistical tension to the intellectual debates between scholars and scientists discussing the nature of scientific research.

The Continuing Debate

The Structure of Scientific Revolutions is now in its fourth edition.

The latest, a 50th anniversary edition, prompted a renewed wave of interest in Kuhn's work.[7] *Structure* continues to provoke and compel those thinking about radical changes in the state of knowledge generally. Even when scholars work in fields far distant from Kuhn's history of science—in history, anthropology, and sociology, for example—they continue to speak in terms of the "structure of revolutions" and "paradigm shifts"* as Kuhn defined the term.[8]

Beyond the academic community, Kuhn himself retains a reputation as a radical opponent of entrenched norms of scientific and academic behavior. In fact, people sometimes see him as a proto-postmodernist—having begun the postmodern interest in including under-represented and marginalized groups in science. Because the reputation of *The Structure of Scientific Revolutions* has widened beyond scientific circles, several terms in the book have entered general use. These include the idea of scientific revolution as an upheaval in scientific knowledge and—most famously—the terms "paradigm" and "paradigm shift." We now hear these equally in popular culture and intellectual environments. Indeed, the expression "paradigm shift" became an Internet buzzword in the late 1990s. Promoters of online shopping used it to convince customers to change their shopping habits.[9] In 2001, *The Complete Idiot's Guide to a Smart Vocabulary* referred to "paradigm shift" as a phrase so overused that it has become meaningless.[10]

1. Carl Hempel, "Thomas Kuhn: Colleague and Friend," in *World Changes: Thomas Kuhn and the Nature of Science*, ed. Paul Horwich (Cambridge, MA: MIT Press, 1993), 7–8.
2. Esther-Mirjam Sent, "Review of Steve Fuller, *Thomas Kuhn: A Philosophical History of our Times*," *The Review of Politics* 63, no. 2 (2001): 392.
3. Paul Gross and Norman Levitt, *Higher Superstition: The Academic Left and Its Quarrels with Science* (Baltimore, MA: Johns Hopkins University Press, 1994).
4. Ziauddin Sardar, "Thomas Kuhn and the Science Wars," in *Postmodernism and Big Science*, ed. Richard Appignanesi (Cambridge: Icon Books, 2002), 189.
5. Peter Godfrey-Smith, *An Introduction to the Philosophy of Science: Theory and Reality* (Chicago, IL: University of Chicago Press, 2003), 146.
6. Michael Bérubé and Cary Nelson, eds, *Higher Education Under Fire: Politics, Economics and the Crisis of the Humanities* (New York: Routledge, 1995), 1.
7. Thomas Nickles, "Introduction," in *Thomas Kuhn*, ed. Thomas Nickles (Cambridge: Cambridge University Press, 2003), 1–19.
8. Carolyn Merchant, "The Theoretical Structure of Ecological Revolutions," in *Out of the Woods*, ed. Char Miller and Hal Rothman (Pittsburgh, PA: Pittsburgh University Press, 2014), 18–27.
9. Kent German, "Top 10 Buzzwords," *CNET*, accessed August 17, 2013, http://www.cnet.com/1990-11136_1-6275610-1.html.
10. Paul McFedries, *The Complete Idiot's Guide to a Smart Vocabulary* (New York: Alpha, 2001), 142–3.

MODULE 12
WHERE NEXT?

KEY POINTS

* *The Structure of Scientific Revolutions* will continue to play a part in debates about how the field of science should respond to attacks on the credibility of its status as an impartial accumulation of knowledge.
* It appears likely that scholars will continue to debate the meaning of Kuhn's work and its significance to the history and philosophy of science.*
* *The Structure of Scientific Revolutions* presented a radical and original understanding of scientific knowledge and practice. It also transformed the public image of scientists and transformed the concept of "paradigm"* into a household term. The work has also inspired generations of scientists and other scholars to reflect on the status of scientific knowledge.

Potential

In the half century since its first publication in 1962, Thomas Kuhn's *The Structure of Scientific Revolutions* has become emblematic of a transformation in the history of science.* As Kuhn himself observed, his work has received credit for more than its share of the revolution in the history of science. This seems likely to continue, as many aspects of the book make it ripe for reinterpretation.[1]

The potential of Kuhn's work lies in its usefulness to the ongoing debate between two schools of contemporary sociology*: the sociology of scientific knowledge* (according to which social conditions create scientific knowledge), and constructivist* sociology (according to which scientists do not study reality directly, they "construct" it from the results of their experiments)—and also to postmodernist* and multicultural* critiques of science. In addition, it continues to challenge the possibility of balancing internalist* readings of science with a view that considers the circumstances in which scientists work.

Kuhn did not intend to spark most of these developments. In fact, he had intended the work for a limited audience of scientists, historians, and philosophers of science. But as long as scholars continue the debate, Kuhn's critique of cumulative, linear, scientific development should remain influential.

We may see Kuhn's book as a source of new ideas that have branched out in ways that he could not have foreseen and may reasonably expect that these branches will continue to flourish and develop. But we may also wonder to what extent these branches will continue to identify themselves as part of Kuhn's legacy. This will also determine whether scholars and practitioners in these fields will go back to the original text in search of new ideas to develop.[2]

> "The purpose of the 'Science Wars'* issue was to answer the 'shrill tone of backlash' against feminist, multiculturalist, and social critics of science ... a backlash designed to intimidate anyone who dares to question the gender-laden assumptions of science, the capitalist foundations of scientific empiricism* and the destructive effects of science and technology on society and environment."
> ——Ziauddin Sadar, *Thomas Kuhn and the Science Wars*

Future Directions

In the absence of a Kuhnian school of thought, those most likely to further the potential of *The Structure of Scientific Revolutions* are the same scholars who responded energetically to Kuhn in the run-up to the publication of the 50th anniversary edition of *The Structure of Scientific Revolutions* in 2012.

These include the American historian and sociologist Steve Fuller* who wrote *Thomas Kuhn: A Philosophical History for Our Time* (2000), a study formed around the central question "how radical were Kuhn's ideas?" His answer is a resounding "not very." *Thomas Kuhn* (2000) by the British philosopher of science Alexander Bird* studies Kuhn as refracted through the central ideas of *Structure*, considered both individually, in the case of "paradigm" and "incommensurability,"* and as a whole in the context of the study of the philosophy of science. The American science philosopher Thomas Nickles* took the opportunity to collect together the latest scholarship on Kuhn's life, times, and work in the edited volume of essays *Thomas Kuhn* (2003). In *Thomas Kuhn's "Linguistic Turn" and the Legacy of Logical Empiricism* (2008), the Italian historian of science Stefano Gattei* focused specifically on the relationship between Kuhn's work and the logical empiricism* it purported to attack.

Kuhn's The Structure of Scientific Revolutions Revisited (2012), a work edited by the Greek historians of science Vasso Kindi* and Theodore Arabatzis,* took stock of recent directions in research prompted by Kuhn's work. In particular, the essays they collected examine the function of concepts in scientific research, logical positivism,* the relationship of history to the philosophy of science, and the nature of progress in science.

As impressive as this list is, these scholars merely study Kuhn's ideas. They do not try to apply them. So Kuhn's work may well remain no more than a chapter of the history of ideas.* As Ian Hacking explained in the anniversary edition of *The Structure of Scientific Revolutions*, "just because *Structure* is a great book, it can be read in endless ways and put to endless uses."[3] Future ideas inspired by Kuhn's work will likely be as much of a surprise to scholarship as *Structure* was when it first appeared in 1962.

Summary

The Structure of Scientific Revolutions deserves to be read for three interconnected reasons:

First, the work transformed our understanding of the nature and characteristics of science. It did so using concepts that have entered everyday language. Perhaps the most widely known of these are the ideas of the paradigm and the paradigm shift.* These concepts, and others discussed in *Structure*, will likely continue to be heard in households and schools and seminar rooms across the world.

Second, Kuhn's work stimulated dramatic changes beyond the history of science, affecting thought in philosophy and a range of social sciences. His notion of scientific revolution has become synonymous with human endeavor and discovery on a grand scale in a way that improves mankind's ability to live and flourish in the world.

Third, the demanding conceptual apparatus Kuhn deploys in his work offers a lesson in critical thought that can help readers of any discipline develop their critical faculties. This is another way in which the work reaches beyond science: even professionals working in business or the knowledge industry can benefit from Kuhn's thinking. In that sense, Kuhn's work may be one of the most broadly useful books of the twentieth century. It seems likely to retain its usefulness as we progress through the twenty-first.

---

1. Thomas S. Kuhn, *The Road Since Structure: Philosophical Essays, 1970—1993 with an Autobiographical Interview*, ed. James Conant and John Haugeland (Chicago, IL: Chicago University Press, 2000), 90–1.
2. Marnie Hughes-Warrington, "Thomas Samuel Kuhn," in *Fifty Key Thinkers in History* (London: Routledge, 2003), 191–2.
3. Ian Hacking, "Introductory Essay," in Thomas S. Kuhn, *The Structure of Scientific Revolutions*, 4th ed. (Chicago, IL: University of Chicago Press, 2012), viii.

GLOSSARY OF TERMS

1. **Analytic proposition:** a statement that is true because of what it means, usually a widely accepted statement of fact:"crows are black," for example.

2. **Analytic-synthetic distinction:** a distinction used by philosophers to separate factual or inherently true statements (analytic propositions) from those rendered true by virtue of what they say about the world (synthetic propositions).

3. **Anomalies:** in the context of Kuhn's concept of paradigm, these are discrepancies between the explanation provided by the paradigm and the aspects of reality it purports to explain. Anomalies cast doubt on paradigms that can lead to periods of crisis science and scientific revolutions.

4. **Classical mechanics:** applied mechanics concerned with the motion and equilibrium of bodies and the action of forces. Branches of classical mechanics include kinematics, dynamics, and statics.

5. **Cognitive science:** the field of study concerned with analyzing how people acquire and use the information they receive through their senses of sight, hearing, taste, and touch.

6. **Cognitive worldview:** the way in which people view the world around them by using their abstract ideas to connect perceptions of reality they gather through their sense of sight, sound, smell, touch, and taste. For example, we know from sight and touch that we stand on the ground as do most objects; we know from the idea of gravity that this is because of gravity, which we cannot otherwise sense.

7. **Cold War:** a period of military tension between the United States and the Soviet Union that lasted from the end of World War II until 1991, when the Soviet Union collapsed.

8. **Consensus paradigm:** the paradigm scientists choose from the competing alternatives during periods of crisis science, the choice creating a scientific revolution.

9. **Constructivist:** an adherent to the school of thought called constructivism in the philosophy of science.

10. **Constructivism:** holds that scientists do not study reality directly but construct

their understanding of it from information they collect through experiments.

11. **Crisis science:** a period of uncertainty that begins when anomalies between reality and a paradigm cast doubt on the consensus paradigm. This period may result in a scientific revolution, when a new consensus paradigm is chosen, or reversion to normal science if anomalies can be explained away with minor amendments to the consensus paradigm.

12. **Darwinian theory:** a theory of the evolution of species, according to which animals and organisms evolve by inheriting from their parents traits in their minds and bodies that increase their chances of competing, surviving, and reproducing.

13. **Disciplinary matrix:** in Kuhn's work, a scientific community's symbolic generalizations, models, values, and problem solutions that they use in their field of specialization.

14. **Empirical:** derived from the study of observable evidence rather than from assumptions or theory.

15. **Empiricism:** the view that all knowledge derives from the experience we gain of reality through our senses (sight, sound, touch, smell, and taste). In science, empiricism therefore emphasizes the importance of accurate experiments in order to "see" reality.

16. **Empiricist:** an adherent of empiricism who, in the context of science, advocates accurate experimentation in order to "see" reality.

17. **Esoteric:** used to describe language, schools of thought or any system of understanding as difficult to understand for an audience or reader with no prior knowledge. In other words, suitable for an inner circle of those in the know.

18. **Exemplar:** a well-known solution to a scientific problem developed under the consensus paradigm during a period of normal science that becomes a model for all scientists working under that paradigm to emulate.

19. **Externalist:** an approach to understanding science as the product of factors external to the scientific community and scientific experiment, such as social factors, political events, and economic determinants.

20. **Extraordinary science:** the period in between crisis science and a scientific revolution when multiple paradigms compete to explain a number of anomalies that have become too great for the existing consensus paradigm to survive. It is out of the ordinary (or "extraordinary") because scientific activity is a state of chaos.

21. **Feminist:** an agenda in politics and scholarship that promotes the equal rights and freedom of women to take any role in life they choose, even those roles traditionally seen as belonging exclusively to the realm of men.

22. **Gestalt psychology:** a school of psychology holding that perceptions to the world are "Gestalts"—thoughts and impressions produced by the mind when it connects perceptions of the world (sights, sounds, language) using theoretical frameworks such as ideas to make sense of them. Gestalt psychology assumes that we build a worldview by mentally ordering our experience in such a way that some individual, new, experience has the capacity to radically alter our understanding of what is "real" altogether.

23. **Gestalt switch:** a moment, according to Gestalt psychology, when an individual's perceptions change from one Gestalt to another. After Gestalt switches, people see the world differently from the way they had before.

24. **Globalization:** the process by which corporations, economies, and nation states have opened their operations and borders to global interaction and communication in the past 30 years, creating new, mass markets.

25. **Harvard University:** a leading private American research university based in Cambridge, Massachusetts, and founded in 1636.

26. **Hiroshima:** a Japanese city in western Honshu, one of two cities on which the US Air Force dropped atomic bombs in 1945, near the end of World War II.

27. **History of ideas:** the sub-discipline of history committed to the study of how people form and change ideas of the world in which they live over time as a function of their social, cultural, economic, political, and religious lives.

28. **History of science:** the sub-discipline of history that focuses on explaining and understanding the development of scientific knowledge over time.

29. **History of Science Society:** the society promotes interest in the history of science and its social and cultural relations across the United States. The presidency of the History of Science Society is an elected administrative office whose candidates are chosen by their peers on merit.
30. **Idealism:** a philosophical school of thought that argues reality is a construction of the human mind rather than something outside ourselves that we can investigate directly.
31. **Incommensurability:** a characteristic of paradigms meaning that they cannot fully understand, or borrow, each other's language, concepts or methodology in any way. Scientists working in different fields, each field having its own paradigm, cannot therefore collaborate easily across paradigms just as scientists in the same field cannot use past paradigms once a new consensus paradigm overturns them.
32. **Internalism:** a position in the philosophy of science that explains science purely in terms of scientists' behaviors and the nature of scientific practice. Internalists do not take external context into account.
33. **Irrationality:** the process of thinking about and reaching conclusions concerning any subject without forming reasonable interpretations of the subject according to widely accepted standards of logic. Contrasts with rationalism.
34. **Logical empiricism:** a movement in the philosophy of science that flourished in the early twentieth century. Logical empiricists insisted that science should analyze parts of reality to understand the whole. It also held that all scientists' conclusions can be verified by having other scientists re-running their experiments.
35. **Logical positivism:** theories and doctrines of the Vienna Circle philosophers in the early 1930s. It proposes that metaphysical, irrational, and speculative questions are logically ill-founded. Instead, logical positivism aims at evolving formal methods—similar to those of the mathematical sciences—to verify empirical questions in the language of philosophy.
36. **Manhattan Project:** a research project that designed and manufactured the first nuclear weapons in the world between 1942 and 1946, led by America but with

the support of Canada and the United Kingdom.

37. **Massachusetts Institute of Technology (MIT):** a private research university based in Cambridge, Massachusetts, founded in 1861, originally with an emphasis on engineering and subjects relating to the technological advancement of the American economy.

38. **Multiculturalism:** policies designed to encourage and nurture to coexistence of different cultures within one society.

39. **Nagasaki:** a Japanese city on the island of Kyushu, one of two cities on which the US Air Force dropped an atomic bomb in 1945, near the end of World War II.

40. **Normal science:** the period during which a consensus paradigm rules the work of scientists in their field, and scientists work to amend the paradigm as their experimentation progresses.

41. **Normative:** a statement that judges or measures according to an ideal standard or norm rather than according to actual reality.

42. **Oil crisis of 1973:** the rise in oil prices from $3 to $12 (US) as a result of an oil export embargo against America, following American intervention in a Middle Eastern war between Israel and a coalition of Arab states led by Egypt and Syria. The crisis led to dramatic and sustained increases in the cost of living in Western nations.

43. **Pallid:** any object or organism lacking depth or intensity of color.

44. **Paradigm:** in Kuhn's understanding of science, a concept agreed by scientists that solves a scientific problem and also guides future scientific research until a paradigm shift.

45. **Paradigm shift:** when a paradigm reaches a state of incommensurability, scientists agree a new paradigm.This means they shift their allegiance from the old paradigm to the new.

46. **Philosophy of science:** the branch of philosophy that seeks to explain the foundational principles by which scientists conduct their work.

47. **Post-colonial studies:** a sub-discipline of the humanities committed to revising

anthropology, history, and sociology (and other disciplines) to include the perspective of the persons and groups previously subjected to colonial rule. Most often, this includes countries in Africa, Asia, South America and elsewhere colonized by European powers.

48. **Postmodernism:** a movement in the humanities and social sciences that took hold by the end of the 1980s. It urged scholars to explore the role that previously unrepresented groups—women, ethnic and religious minorities, and marginalized sexualities—had played in creating the past.

49. **Pre-consensus science:** the earliest phases of scientific activity when scientists from rival schools compete to interpret phenomena. This competition is resolved when scientists choose the first paradigm according to which they all work, and the first period of normal science ensues.

50. **Princeton University:** a leading private research university in Princeton, New Jersey, founded in 1746.

51. **Quantum mechanics:** the branch of physics that studies the motion and interaction of the smallest particles. Quantum mechanics deepens the findings of classical mechanics by focusing on a smaller scale in the physical world.

52. **Quantum physics:** a branch of physics studying the smallest level of objects in the physical universe: atoms and atomic particles.

53. **Rationalism:** the philosophical school of thought that attaches importance to the use of logic and reasoned thought to solve problems rather than relying on intuition or faith.

54. **Realism:** the school of thought in the philosophy of science that argues scientists can access and describe reality directly without their own opinions and practices affecting the truth of their conclusions.

55. **Relativism:** a school of thought in science which holds that scientists reach provisional conclusions rather than absolute definitive results that will never change when they conduct their research.

56. **Revolutionary science:** the period during which several paradigms compete to become the consensus paradigm of a community of scientists, the original

paradigm having become doubtful after reality throws up anomalies that it cannot explain.

57. **Science wars:** a series of intellectual exchanges in the 1990s and early 2000s when scientific realists and postmodernist critics debated the nature of science.

58. **Scientific community:** all the scientists working in the world at any one time.

59. **Scientific revolution:** the process occurring from time to time that causes the development of scientific knowledge, usually following new discoveries. During a revolution, scientists discard an old paradigm, starting instead to work in accordance with a new one that better explains the reality they investigate.

60. **Sociology:** the study of the structure and history of human societies.

61. **Sociology of science:** the branch of sociology that examines how scientists' social, economic, and professional positions help shape the research topics and conclusions produced in the scientific community.

62. **Sociology of scientific knowledge (SSK):** a school of thought within the sociology of science that takes inspiration from Kuhn's work and argues for a constructivist interpretation of science.

63. **Soviet Union:** the Union of Soviet Socialist Republics that existed from 1922 until 1991. Its center of power was Russia, but it also encompassed many other states, including Ukraine and Georgia.

64. **Synthetic proposition:** a statement that is true because of the way in which it relates to the world rather than because it is a statement of fact—for example,"crows attack small birds."The statement is not true of crows in general, but is borne out by the observable behavior of a majority of crows.

65. **Tacit knowledge:** knowledge not easily transferred from one person to another by the simple act of writing or speaking, including assumptions often taken almost for granted.

66. **Teleology:** the study of a subject such as the history of science as leading inevitably to an end or final cause. Teleology often comes in for criticism for ignoring the chaotic journey of science to its current form owing to excessive focus on the destination rather than the journey.

67. **Theoretical physics:** the branch of physics that relies on theory, mathematics, and abstract reasoning in order to present possible explanations of the natural world that could not be tested in a laboratory due to the size or nature of the subject considered. For example, one could not perform laboratory tests on the nature of the universe.

68. **Unit ideas:** the basic ideas and perceptions of the world that make no sense on their own until joined together by the thinking mind into a worldview, as explained by Arthur O. Lovejoy. For example, the idea that "the sky is blue" joins together unit ideas including being able to see a sky, being able to distinguish blue from other colors, and knowing the words "sky" and "blue."

69. **Verifiability theory of meaning:** a theory asserting that the meaning and truth of statements or scientific conclusions lies in the ability of a scientist to re-create an experiment conducted by her colleague and still reach the same conclusions, so checking its truth. If different conclusions result from the same experiment then the scientist has shown her colleague's finding to be false.

70. **Verificationist principle:** the rule that the conclusions of scientific experiment must be replicable by third parties when they repeat the same experiment in order for them to be true.

71. **University of California, Berkeley:** the leading research university in the University of California system and a major public-research university with an international reputation for its teaching and research.

72. **World War I (1914–18):** the global armed conflict between the Allied powers (Britain, France, Russia, and the United States) on the one side and the Central Powers (Germany, the Ottoman Empire and Austria-Hungary) on the other.

PEOPLE MENTIONED IN THE TEXT

1. **Hanne Andersen** is a Danish philosopher of science in the Faculty of Technology and Science, Department of Mathematics and Science Studies at Aarhus University. She has studied Kuhn's work extensively, and works to make the philosophy of science relevant to scientific education.

2. **Theodore Arabatzis** is an associate professor of philosophy and history of science at the University of Athens, Greece. He has recently published *Representing Electrons* (2006).

3. **Francis Bacon (1561–1626)** was an English philosopher, statesman, lawyer, and writer. His main works are *Novum Organum Scientiarum* (1620) and *New Atlantis* (1627).

4. **Peter Barker** is a British historian and philosopher of science who currently holds the appointment of professor of history of science at the University of Oklahoma.

5. **Barry Barnes (b. 1943)** is a British philosopher of science, currently professor of sociology at the University of Exeter. He worked with David Bloor from the 1970s to the 1990s at the University of Edinburgh to develop the strong program of the sociology of scientific knowledge.

6. **George Berkeley (1685–1753)** was a British philosopher and one of the principal exponents of the British empiricist philosophy, according to which knowledge derives from experience, not reason. His main work is *A Treatise on the Principles of Human Knowledge* (1710).

7. **Michael Bérubé (b. 1961)** is an American philosopher and Professor and Director of the Institute for the Arts and Humanities at Pennsylvania State University, USA. He has written *Higher Education Under Fire* (1995).

8. **Alexander Bird** is a British philosopher of science and professor of philosophy at the University of Bristol. He wrote *Thomas Kuhn* (2000) and *Philosophy of Science* (1998).

9. **David Bloor (b. 1942)** is a British sociologist and professor of science and technology studies at the University of Edinburgh. He is best known for having established the strong program of sociology of scientific knowledge, particularly

in his work *Knowledge and Social Imagery* (1978).

10. **J. Z. Buchwald (b. 1949)** is an American historian of science who is currently Doris and Henry Dreyfuss Professor of History at the California Institute of Technology at Pasadena, California. He has written on a variety of history of science topics including an edited volume on the practice of physics, *Scientific Practice:Theories and Stories of Doing Physics* (1995).

11. **Rudolf Carnap (1891–1970)** was a German logical-empiricist philosopher of science. He emigrated to the United States in 1935 and his main work is *Philosophy and Logical Syntax* (1935).

12. **Xiang Chen** is an Asian American philosopher of science, currently professor of philosophy at the California Lutheran University.The main contribution on the subject is Andersen, Barker and Chen, *The Cognitive Structure of Scientific Revolutions* (2006).

13. **James Bryant Conant (1893–1978)** was an American chemist, president of Harvard University and the first US ambassador to West Germany.

14. **Nicolaus Copernicus (1473–1543)** was a Polish Renaissance mathematician and astronomer who formulated the current idea that the sun is at the center of the universe and so the earth rotates around it.

15. **Arthur Danto (1924–2013)** was an American art critic and philosopher who contributed to a number of fields, notably historical theory. He is most widely remembered as an art critic for *The Nation*, the USA's oldest weekly magazine.

16. **Charles Darwin (1809–82)** was an English naturalist who proposed that all species of life have descended from common ancestors, and that evolution resulted from natural selection, or the survival of the fittest.

17. **Pierre Duhem (1861–1916)** was a French historian and philosopher of science. His masterpiece is *The System of theWorld* (1913–16), in which he famously argued for the continuation between medieval and early modern science.

18. **Paul Feyerabend (1924–94)** was an Austrian-born émigré to the United States and iconoclast philosopher of science. His main work is *Against Method* (1975).

19. **Michel Foucault (1926–84)** was a French philosopher, historian, social theorist,

philologist, and psychologist, one of the twentieth century's most famous thinkers. His main work is *The Archaeology of Knowledge* (1969).

20. **Michael Friedman (b. 1947)** is an American philosopher of science who teaches and researches at Stanford University. His book *Dynamics of Reason* (2001) developed and explored areas of Kuhn's concept of paradigm shifts that Kuhn himself left under-developed.

21. **Steve Fuller (b. 1959)** is an American philosopher and sociologist. His main works are *Kuhn vs Popper* (2003) and *Science vs Religion?* (2007).

22. **Stefano Gattei** is an Italian philosopher of science who is a member of the faculty at the Institute for Advanced Studies, Lucca, Italy. His doctoral dissertation studied Kuhn's work: *La Rivoluzione Incompiuta di Thomas Kuhn* ("Thomas Kuhn's Incomplete Revolution") (2007).

23. **Jan Golinski (b. 1957)** is an American philosopher of science and a professor of history and humanities at the University of New Hampshire. He is the author of *Making Natural Knowledge* (1998).

24. **Paul Gross** is an American biologist and author of *Higher Superstition* (1994) with Norman Levitt, as well as *Creationism's Trojan Horse: The Wedge of Intelligent Design* (2004).

25. **Ian Hacking (b. 1936)** is an influential Canadian historian and philosopher of science. His main work is *The Emergence of Probability* (1975).

26. **Norwood Russell Hanson (1924–67)** was an American philosopher of science. He is best-known for arguing that when scientists observe the world they impose a host of preconceived theories on what they see.

27. **John L. Heilbron (b. 1934)** is an American historian of science, and professor of history at the University of California at Berkeley. His main work is *Electricity in the 17th and 18th Centuries: A Study of Early Modern Physics* (1979).

28. **Carl Hempel (1905–97)** was a German American philosopher of science and famous logical empiricist. His main work is *Scientific Explanation* (1967).

29. **Richard Hofstadter (1916–70)** was an American historian of American history in the modern and contemporary era. His work ranged widely across the history

of ideas and social history more generally.

30. **Paul Hoyningen-Huene (b. 1946)** is a German philosopher of science best known for his neo-Kantian interpretation of Thomas Kuhn. His main work is *Reconstructing Scientific Revolutions: Thomas S. Kuhn's Philosophy of Science* (1993).

31. **David Hume (1711–76)** was a British philosopher, one of the principal exponents of British empiricist philosophy, according to which knowledge derives from experience, not reason. His main writing is *A Treatise of Human Nature* (1739).

32. **Immanuel Kant (1724–1804)** was a German philosopher and one of the most important thinkers of the modern age. His major work is *The Critique of Pure Reason* (1781).

33. **Johannes Kepler (1571–1630)** was a German mathematician and astronomer who played an important role in revising scientists' understanding of the world in the seventeenth century. His theory of the arrangement of the planets that gives us our current model of the universe forms his main contribution.

34. **Vasso Kindi** is a Greek philosopher of science who is assistant professor of philosophy and history of science at the University of Athens, Greece. Her main work so far is *Kuhn & Wittgenstein: Philosophical Investigation of the Structure of Scientific Revolutions* (1995).

35. **Alexandre Koyré (1892–1964)** was a French Russian émigré to the USA, who developed outstanding analyses of seventeenth-century history of science and helped coin the term "scientific revolution" to describe the intellectual changes of the period. His main work is *From the Closed World to the Infinite Universe* (1957).

36. **Imre Lakatos (1922–74)** was a Hungarian philosopher of science and mathematician. His main work is *The Methodology of Scientific Research Programmes* (1978).

37. **Bruno Latour (b. 1947)** is a French sociologist of science and anthropology. His main works, *Science in Action* (1987) and *Laboratory Life* (1979), are

famous for reducing scientists' work to the results of their social circumstances.

38. **Antoine Lavoisier (1743–94)** was a French chemist widely considered the "father of modern chemistry" and the leader of the "chemical revolution" of the late eighteenth century.

39. **Norman Levitt (1943–2009)** was an American mathematician at Rutgers University, USA. He is mainly remembered for his book *Higher Superstition* (1994), co-authored with Paul Gross.

40. **John Locke (1632–1704)** was an English philosopher considered a British empiricist and the father of classical liberalism. His main work is *An Essay of Human Understanding* (1690).

41. **Arthur O. Lovejoy (1873–1962)** was a German-born American émigré renowned for his book *The Great Chain of Being* (1936) that began the modern study of the history of ideas.*

42. **Ernst Mach (1838–1916)** was a German philosopher and author of *The Science of Mechanics* (1883), which emphasized empiricism over metaphysics.

43. **Anneliese Maier (1905–71)** was a German historian of science. A selection of her writings was published in English as *On the Threshold of Exact Science: Selected Writings of Anneliese Maier on Late Medieval Natural Philosophy* (1982).

44. **James A. Marcum** is an American philosopher of science who works at Baylor University, Waco, Texas. He has studied the development of science and written most recently *The Conceptual Foundations of Systems Biology: An Introduction* (2009).

45. **Margaret Masterman (1910–86)** was a British linguist and philosopher, best known for her pioneering studies of computational linguistics and automatic computer translation.

46. **Hélène Metzger (1889–1944)** was an influential French historian and philosopher of science. Her main works are *Les Doctrines Chimiques en France du Début du XVIIe à la Fin du XVIIIe siècle* ("The Doctrines of Chemistry in France from the Beginning of the Seventeenth until the End of the Eighteenth

Century") (1923).

47. **Alan Musgrave (b. 1940)** is a New Zealand philosopher of science. He was a student of Karl Popper and author of *Common Sense, Science and Scepticism* (1992).

48. **Otto Neurath (1882–1945)** was an Austrian logical-positivist philosopher and leader of theVienna Circle of philosophy of science. Neurath's main English-language book is *Empiricism and Sociology* (1973).

49. **Isaac Newton (1642–1727)** was an English physicist and mathematician whose work pioneered the laws of motion and the modern theory of gravity. He is best known for his *Principia Mathematica* ("Mathematical Principles") (1687).

50. **Thomas Nickles** is an American philosopher of science who is Foundation Professor at the University of Nevada, Reno. He has written widely on Kuhn as well as other areas of the history of science.

51. **George Orwell (1903–50)** is the pen name of Eric Arthur Blair, an English novelist and journalist. He is famous for his dystopian novel *Nineteen Eighty-Four* (1949) and his novella *Animal Farm* (1945), both of which condemned overbearing governments.

52. **Jean Piaget (1896–1980)** was a Swiss psychologist and philosopher who propounded a famous theory of cognitive development in children. His main works are *The Origins of Intelligence in Children* (1953) and The *Child's Construction of Reality* (1955).

53. **Max Planck (1858–1947)** was a German theoretical physicist who is deemed the founder of quantum theory, which won him the Nobel Prize in Physics in 1918.

54. **Michael Polanyi (1891–1976)** was a British Hungarian philosopher of science and a chemist. His most famous works are *Personal Knowledge* (1958) and *The Tacit Dimension* (1966).

55. **Karl Popper (1902–94)** was a British Austrian philosopher of science, widely considered one of the greatest of the twentieth century. His most famous work is *The Logic of Scientific Discovery* (1934).

56. **Willard Van Orman Quine (1908–2000)** was an influential American philosopher and logician. His main work is *Word and Object* (1960).

57. **Jerry Ravetz (b. 1929)** is an American philosopher of science and environment consultant who is best known for his work on the uncertainty and the ethics of scientific research. His book *A No-Nonsense Guide to Science* (2005) is his best-known contribution in this direction.

58. **George A. Reisch** is an American historian of science best known for his externalist studies of science, for example, *How the Cold War Transformed the Philosophy of Science* (2005).

59. **Howard Sankey** is an Australian philosopher of science who is currently associate professor of philosophy at the University of Melbourne. He is the author of *The Incommensurability Thesis* (1993).

60. **Ziauddin Sardar (b. 1951)** is a well-known British writer, scholar, and public figure. He has written numerous books, including *Reading the Qur'an* (2011) and *Thomas Kuhn and the ScienceWars* (2000).

61. **George Sarton (1884–1956)** was a Belgian-born American chemist and historian of science. He intended to write a nine-volume history of science but at the time of his death had completed only three.

62. **Dudley Shapere** was an American philosopher of science and professor at Wake Forest University. He is mainly remembered for his harsh criticism of Thomas Kuhn and Paul Feyerabend.

63. **Alan Sokal (b. 1955)** is professor of mathematics at University College London and professor of physics at NewYork University.

64. **N. M. Swerdlow (b. 1941)** is professor emeritus of history, astronomy, and astrophysics at the University of Chicago, and visiting professor at the California Institute of Technology.

65. **Harry S.Truman (1884–1972)** was an American politician and 33rd president of the United States from 1945–53, the period immediately after World War II during which the Cold War began.

66. **E. O.Wilson (b. 1929)** is an American biologist, naturalist, and author, who

is the world's leading expert on ants and a retired Harvard professor. His most recent work is *The Meaning of Human Existence* (2014).

67. **Norton Wise (b. 1940)** is Distinguished Professor of History at the University of California at Los Angeles. His research focuses on science and industrialization from the eighteenth century to the present.

68. **Ludwig Wittgenstein (1889–1961)** is often considered the greatest philosopher of the twentieth century; his main works are *Tractatus Logico-Philosophicus* (1921) and *Philosophical Investigations* (1953).

69. **Stephen Woolgar (b. 1950)** is a British sociologist, currently Head of Science and Technology Studies at the Said Business School, University of Oxford. He co-authored *Laboratory Life* (1979) with Bruno Latour.*

70. **John Worrall (b. 1946)** is professor of philosophy of science at the London School of Economics. He was a student of Imre Lakatos and is the author of *The Ontology of Science* (1994).

71. **John M. Ziman (1925–2005)** was a British-born physicist based in New Zealand whose work proved profession-leading in condensed-matter physics. His widest reputation outside science is as a spokesperson for science, and *Real Science* (2000) is his best-known publication on that topic.

WORKS CITED

1. Andersen, Hanne, Peter Barker and Xiang Chen. *The Cognitive Structure of Scientific Revolutions*. Cambridge: Cambridge University Press, 2006.
2. "Kuhn's Mature Philosophy of Science and Cognitive Psychology." *Philosophical Psychology* 9 (1996): 347–63.
3. Barker, Peter, Xiang Chen and Hanne Andersen. "Kuhn on Concepts and Categorization." In *Thomas Kuhn*, edited by Thomas Nickles, 212–45. Cambridge: Cambridge University Press, 2003.
4. Barnes, Barry. *T.S. Kuhn and Social Science*. London: Macmillan, 1982.
5. Bérubé, Michael. "The Science Wars Redux." *Democracy Journal* 19 (2011): 66–74.
6. Bérubé, Michael and Cary Nelson, eds. *Higher Education Under Fire: Politics, Economics and the Crisis of the Humanities*. New York: Routledge, 1995.
7. Bird, Alexander. "The Structure of Scientific Revolutions: An Essay Review of the Fiftieth Anniversary Edition." *The British Journal for the Philosophy of Science* 63, no. 4 (2012): 859–83.
8. *Thomas Kuhn*. Chesham: Acumen, 2000.
9. "Thomas Kuhn." *The Stanford Encyclopedia of Philosophy* (Winter 2014 Edition). Edited by Edward N. Zalta. Accessed July 8, 2015. http://plato.stanford.edu/archives/win2012/entries/davidson/.
10. Bloor, David. *Knowledge and Social Imagery*. 2nd ed. Chicago, IL: University of Chicago Press, 1991.
11. Chang, Hasok. "Incommensurability: Revisiting the Chemical Revolution." In *Kuhn's The Structure of Scientific Revolutions Revisited*, edited by Vasso Kindi and Theodore Arabatzis, 153–79. New York: Routledge, 2012.
12. Curd, M.V. "Kuhn, Scientific Revolutions and the Copernican Revolution." *Nature and System* 6 (1984): 1–14.
13. Danto, Arthur. *Narration and Knowledge*. New York: Columbia University Press, 1985.
14. Feyerabend, Paul. "Consolations for the Specialist." In *Criticism and the Growth of Knowledge*, edited by Imre Lakatos and A. Musgrave, 197–229. Cambridge:

Cambridge University Press, 1970.

15. Friedman, Michael. "Kuhn and Logical Empiricism." In *Thomas Kuhn*, edited by Thomas Nickles, 19–44. Cambridge: Cambridge University Press, 2003.

16. "Remarks on the History of Science and the History of Philosophy." In *World Changes: Thomas Kuhn and the Nature of Science*, edited by Paul Horwich, 37–54. Cambridge, MA: MIT Press, 1993.

17. Fuller, Steve. *Kuhn vs Popper: The Struggle for the Soul of Science*. New York: Columbia University Press, 2004.

18. *Thomas Kuhn: A Philosophical History for Our Times*. Chicago, IL: University of Chicago Press, 2000.

19. Gattei, Stefano. *Thomas Kuhn's "Linguistic Turn" and the Legacy of Logical Empiricism*. Burlington, VT: Ashgate, 2008.

20. German, Kent. "Top 10 Buzzwords," *CNET*. Accessed August 17, 2013. http://The Structure of Scientific Revolutions.cnet.com/1990-11136_1-6275610-1. html.

21. Godfrey-Smith, Peter. *An Introduction to the Philosophy of Science: Theory and Reality*. Chicago, IL: University of Chicago Press, 2003.

22. Green, Christopher. "Where Is Kuhn Going?" *American Psychologist* 59, no. 4 (2004): 271–72.

23. Gross, Paul, and Norman Levitt. *Higher Superstition: The Academic Left and Its Quarrels with Science*. Baltimore, MD: Johns Hopkins University Press, 1994.

24. Hacking, Ian. "Introductory Essay." In Thomas Kuhn, *The Structure of Scientific Revolutions*, 4th edition, vii–xxxvii Chicago, IL: University of Chicago Press, 2012.

25. *Scientific Revolutions*. Oxford: Oxford University Press, 1981.

26. *The Social Construction of What?* Cambridge, MA: Harvard University Press, 1999.

27. Hall, A. Rupert. *The Revolution in Science 1500–1750*. London: Longman, 1983.

28. Hall, Marie Boas. "Review of *The Structure of Scientific Revolutions*." *American Historical Review* 68, no. 3 (1963): 700–1.

29. Heilbron, John L. "A Mathematicians' Mutiny, with Morals." In *World Changes: Thomas Kuhn and the Nature of Science*, edited by Paul Horwich, 81–129. Cambridge, MA: MIT Press, 1993.
30. "Thomas Samuel Kuhn." *Isis* 89, no. 3 (1998): 505–15.
31. Hempel, Carl. "Thomas Kuhn: Colleague and Friend." In *World Changes: Thomas Kuhn and the Nature of Science*, edited by Paul Horwich, 7–8. Cambridge, MA: MIT Press, 1993.
32. Hoftstadter, Richard. *Social Darwinism in American Thought, 1860–1915*. Boston, MA: Beacon, 1944.
33. Horwich, Paul. "Introduction." In *World Changes: Thomas Kuhn and the Nature of Science*, edited by Paul Horwich, 1–5. Cambridge, MA: MIT Press, 1993.
34. Hoyningen-Huene, Paul. *Reconstructing Scientific Revolutions: Thomas S. Kuhn's Philosophy of Science*. Chicago, IL: University of Chicago Press, 1993.
35. "Two Letters of Paul Feyerabend to Thomas S. Kuhn on a Draft of The Structure of Scientific Revolutions." *Studies in History and Philosophy of Science* 26, no. 3 (1995): 353–88.

36. Hughes, Jeff. "Whigs, Prigs and Politics: Problems in the Historiography of Contemporary Science." In *The Historiography of Contemporary Science and Technology*, edited by Thomas Söderqvist, 19–39. Amsterdam: Harwood Academic Publishers, 1997.
37. Hughes-Warrington, Marnie. *Fifty Key Thinkers on History*. London: Routledge, 2003.
38. Irzik, Gurol, and Teo Grünberg. "Carnap and Kuhn: Arch Enemies or Close Allies?" *British Journal for the Philosophy of Science* 46, no. 3 (1995): 285–307.
39. Jewett, Andrew. *Science, Democracy and the American University: From the Civil War to the Cold War*. Cambridge: University of Cambridge Press, 2012.
40. Kindi, Vasso, and Theodore Arabatzis, "Introduction." In *Kuhn's The Structure of Scientific Revolutions Revisited*, edited by Vasso Kindi and Theodore Arabatzis, 1–15. New York: Routledge, 2012.
41. Kuhn, Thomas S. "Afterwords." In *World Changes: Thomas Kuhn and the Nature of Science*, edited by Paul Horwich, 311–43. Cambridge, MA: MIT Press,

1993.

42. *Black Body Theory and Quantum Discontinuity, 1894–1912*. Chicago, IL: Chicago University Press, 1978.

43. "Concepts of Cause in the Development of Physics." In *The Essential Tension: Selected Studies in Scientific Tradition and Change*, 21–31. Chicago, IL: University of Chicago Press, 1977.

44. *The Copernican Revolution: Planetary Astronomy in the Development of Western Thought*. Cambridge, MA: Harvard University Press, 1957.

45. "The Function of Dogma in Scientific Research." In *Scientific Change*, edited by Alistair C. Crombie, 347–69. New York: Basic Books, 1963.

46. "The Function of Measurement in Modern Physical Sciences." In *The Essential Tension: Selected Studies in Scientific Tradition and Change*, 178–225. Chicago, IL: University of Chicago Press, 1977.

47. "Logic of Discovery or Psychology of Research?" In *Criticism and the Growth of Knowledge: Proceedings of the International Colloquium in the Philosophy of Science*, edited by Imre Lakatos and A. Musgrave, 1–23. Cambridge: Cambridge University Press, 1970.

48. "Objectivity, Value Judgement and Theory Choice." In *The Essential Tension: Selected Studies in Scientific Tradition and Change*, edited by Thomas S. Kuhn, 320–39. Chicago, IL: Chicago University Press, 1977.

49. "The Relations between the History and the Philosophy of Science," in *The Essential Tension: Selected Studies in Scientific Tradition and Change*, 1–21. Chicago, IL: University of Chicago Press, 1977.

50. *The Road Since Structure: Philosophical Essays, 1970–1993 with an Autobiographical Interview*. Edited by James Conant and John Haugeland. Chicago, IL: Chicago University Press, 2000.

51. "Second Thoughts on Paradigms." In *The Essential Tension: Selected Studies in Scientific Tradition and Change*, 295–319. Chicago, IL: University of Chicago Press, 1977.

52. *The Structure of Scientific Revolutions*, 4th edition. Chicago, IL: University of Chicago Press, 2012.

53. "Theory-Change as Structure-Change: Comments on the Sneed Formalism." *Erkenntnis* 10 (1976): 179–99.
54. Lakatos, Imre. "Falsification and the Methodology of Scientific Research Programmes." In *Criticism and the Growth of Knowledge*, edited by Imre Lakatos and A. Musgrave, 91–195. Cambridge: Cambridge University Press, 1970.
55. Latour, Bruno and Steve Woolgar. *Laboratory Life: The Construction of Scientific Facts*. New York: Sage Publications, 1979.
56. Longino, Helen E. "Does the Structure *of Scientific Revolutions* Permit a Feminist Revolution in Science?" In *Thomas Kuhn*, edited by Thomas Nickles, 261–81. Cambridge: Cambridge University Press, 2003.
57. Mandelbaum, Maurice. "A Note on Thomas S. Kuhn's *The Structure of Scientific Revolutions*." *The Monist* 60, no. 4 (1977): 445–52.
58. Marcum, James A. *Thomas Kuhn's Revolution: An Historical Philosophy of Science*. London: Continuum, 2005.

59. Margolis, Joseph. "Objectivity as a Problem." *Annals of the American Academy of Political and Social Science* 560 (1998): 55–68.
60. Martin, R. N. D. *Pierre Duhem: Philosophy and History in the Work of a Believing Physicist*. La Salle, IL: Open Court, 1991.
61. Massimi, Michela. "Philosophy and the Sciences After Kant." In *Conceptions of Philosophy*, edited by Anthony O'Hear, 275–312. Cambridge: Cambridge University Press, 2000.
62. Masterman, Margaret. "The Nature of a Paradigm." In *Criticism and the Growth of Knowledge*, edited by Imre Lakatos and A. Musgrave, 59–89. Cambridge: Cambridge University Press, 1970.
63. McFedries, Paul. *The Complete Idiot's Guide to a Smart Vocabulary*. New York: Alpha, 2001.
64. Merchant, Carolyn. "The Theoretical Structure of Ecological Revolutions." In *Out of the Woods*, edited by Char Miller and Hal Rothman, 18–27. Pittsburgh, PA: Pittsburgh University Press, 2014.
65. Musgrave, Alan. "Kuhn's Second Thoughts." *British Journal for the Philosophy*

of Science 22, no. 3 (1971): 287–97.

66. Nersessian, Nancy. "Kuhn, Conceptual Change, and Cognitive Science." In *Thomas Kuhn*, edited by Thomas Nickles, 178–211. Cambridge: Cambridge University Press, 2003.

67. Naughton, John. "Thomas Kuhn: The Man Who Changed the Way the World Looked at Science." *Guardian*, August 19, 2012. Accessed July 10, 2015. http://www.theguardian.com/science/2012/aug/19/thomas-kuhn-structure-scientific-revolutions.

68. Newton-Smith, W. H. *The Rationality of Science*. London: Routledge, 1981.

69. Nickles, Thomas. "Introduction." In *Thomas Kuhn*, edited by Thomas Nickles, 1–19. Cambridge: Cambridge University Press, 2003.

70. Novick, Peter. *That Noble Dream: The "Objectivity Question" and the American Historical Profession*. Cambridge: Cambridge University Press, 1988.

71. Popper, Karl. "Normal Science and Its Dangers." In *Criticism and the Growth of Knowledge*, edited by Imre Lakatos and A. Musgrave, 51–8. Cambridge: Cambridge University Press, 1970.

72. Putnam, Hilary. *Reason, Truth and History*. Cambridge: Cambridge University Press, 1981.

73. Reeves, Thomas C. *Twentieth-Century America: A Brief History*. Oxford: Oxford University Press, 2000.

74. Reisch, George A. "Did Kuhn Kill Logical Empiricism?" *Philosophy of Science* 58 (1991): 264–77.

75. Rouse, Joseph. "Kuhn's Philosophy of Scientific Practice." In *Thomas Kuhn*, edited by Thomas Nickles, 100–21. Cambridge: Cambridge University Press, 2003.

76. Sankey, Howard. "Kuhn's Changing Concept of Incommensurability." *British Journal for the Philosophy of Science* 44, no. 4 (1993): 759–74.

77. Sardar, Ziauddin. "Thomas Kuhn and the Science Wars." In *Postmodernism and Big Science*, edited by Richard Appignanesi, 189–228. Cambridge: Icon Books, 2002.

78. Sarkar, Sahotra. *The Legacy of the Vienna Circle: Modern Reappraisals*. New York: Garland Publishing, 1996.
79. Sarton, George. *A Guide to the History of Science*. Waltham, MA: Chronica Botanica Co, 1952.
80. Sent, Esther-Mirjam. "Review of Steve Fuller, *Thomas Kuhn: A Philosophical History of our Times*." *The Review of Politics* 63, no. 2 (2001): 390–2.
81. Shapere, Dudley. "The Paradigm Concept." *Science* 172 (1971): 706–9.
82. "The Structure of Scientific Revolutions." *Philosophical Review* 73, no. 3 (1964): 383–94.
83. Sokal, Alan. "A Physicist Experiments with Cultural Studies." *Lingua Franca* (1996): 62–4.
84. Swerdlow, N. M. "Thomas S. Kuhn, A Biographical Memoir." *National Academy of Science* (2013). Accessed June 29, 2015, http://www.nasonline.org/publications/biographical-memoirs/memoir-pdfs/kuhn-thomas.pdf.

85. "Science and Humanism in the Renaissance: Regiomontanus's Oration on the Dignity and Unity of the Mathematical Sciences." In *World Changes: Thomas Kuhn and the Nature of Science*, edited by Paul Horwich, 131–68. Cambridge, MA: MIT Press, 1993.
86. Watkins, J. "Against 'Normal Science.'" In *Criticism and the Growth of Knowledge*, edited by Imre Lakatos and A. Musgrave, 25–37. Cambridge: Cambridge University Press, 1970.
87. Wise, Norton M. "Mediations: Enlightenment Balancing Acts, or the Technologies of Rationalism." In *World Changes: Thomas Kuhn and the Nature of Science*, edited by Paul Horwich, 207–56. Cambridge, MA: MIT Press, 1993.
88. Wittgenstein, Ludwig. *Philosophical Investigations*, 3rd ed. Oxford: Blackwell, 2003.
89. Worrall, John. "Normal Science and Dogmatism, Paradigms and Progress: Kuhn 'Versus' Popper and Lakatos." In *Thomas Kuhn*, edited by Thomas Nickles, 65–100. Cambridge: Cambridge University Press, 2003.
90. Ziman, John M. *Real Science: What It Is and What it Means*. Cambridge: Cambridge University Press, 2000.

原书作者简介

托马斯·库恩（Thomas Kuhn）于1922年出生在美国俄亥俄州辛辛那提市，18岁时就立志当科学家。他在哈佛大学学习物理学，并获得博士学位。其间学术休假，并在第二次世界大战期间服兵役。随后，库恩做了三年的哈佛初级研究员，他充分享受学术自由，充分利用这一优势涉猎历史和科学哲学——这是他随后取得成功的领域。库恩于1996年去世，其成就不仅挑战了他的学术同行，而且改变了大众思考各个学科的方式。

本书作者简介

乔·赫德桑博士（Dr Jo Hedesan）在牛津大学任教，并担任威康信托研究员，致力于该大学病史与人文学科领域的研究。

约瑟夫·滕德勒博士（Dr Joseph Tendler）获得圣安德鲁斯大学博士学位，是历史编纂学专家，《安纳莱斯学派的反对者》的作者。

世界名著中的批判性思维

《世界思想宝库钥匙丛书》致力于深入浅出地阐释全世界著名思想家的观点，不论是谁、在何处都能了解到，从而推进批判性思维发展。

《世界思想宝库钥匙丛书》与世界顶尖大学的一流学者合作，为一系列学科中最有影响的著作推出新的分析文本，介绍其观点和影响。在这一不断扩展的系列中，每种选入的著作都代表了历经时间考验的思想典范。通过为这些著作提供必要背景、揭示原作者的学术渊源以及说明这些著作所产生的影响，本系列图书希望让读者以新视角看待这些划时代的经典之作。读者应学会思考、运用并挑战这些著作中的观点，而不是简单接受它们。

ABOUT THE AUTHOR OF THE ORIGINAL WORK

Born in Cincinnati, Ohio, in 1922, Thomas Kuhn knew by the age of 18 that he wanted to be a scientist. He studied physics at Harvard University, obtaining his PhD there after a break with academia to serve in the military during World War II. Three years as a Junior Harvard Fellow then gave Kuhn academic freedom, and he took full advantage by moving towards history and the philosophy of science—areas where he would go on to make his mark. Kuhn died in 1996, having both challenged his academic colleagues and changed the way the general public thinks about an entire range of disciplines.

ABOUT THE AUTHORS OF THE ANALYSIS

Dr Jo Hedesan lectures at Oxford and is a Wellcome Trust Research Fellow in Medical History and Humanities at the university.

Dr Joseph Tendler received his PhD from the University of St Andrews. He is a specialist in historiography, the study of how history is conceived and written, and is the author of *Opponents of the Annales School*.

ABOUT MACAT
GREAT WORKS FOR CRITICAL THINKING

Macat is focused on making the ideas of the world's great thinkers accessible and comprehensible to everybody, everywhere, in ways that promote the development of enhanced critical thinking skills.

It works with leading academics from the world's top universities to produce new analyses that focus on the ideas and the impact of the most influential works ever written across a wide variety of academic disciplines. Each of the works that sit at the heart of its growing library is an enduring example of great thinking. But by setting them in context — and looking at the influences that shaped their authors, as well as the responses they provoked — Macat encourages readers to look at these classics and game-changers with fresh eyes. Readers learn to think, engage and challenge their ideas, rather than simply accepting them.

批判性思维与《科学革命的结构》

首要批判性思维技巧：创造性思维

次要批判性思维技巧：推理

毫不夸张地说，托马斯·库恩的《科学革命的结构》可以看作思想史上的里程碑文献。

库恩在解析科学思维的转换时，质疑了当时盛行的观点：科学是走向真理的不间断进程。他认为，科学通过"范式转移"实际取得了进步，这意味着现存科学模式存在瑕疵的证据日渐增多。这首先面对的是反对与疑问，最终导致一场危机，促使一种新模式的发展。这种发展反过来产生一段瞬息万变的时期——库恩称之为"非常规科学"——在最终回归"常规科学"之前，这个过程就开始了，并由此整个周期最终自行循环往复。

把科学描绘成连续革命的产物，是严谨而富有想象力的批判性思维的产物。它与科学的自我形象不一致，即科学作为一套准则，基于现有知识的发展不断演变和改进。库恩对这一形象富有高度创造性的重塑，证明了其具有持久的影响力——是作者能力的直接产物，即对现存证据进行新解释和重新定义，以便以新的方式看待这些问题。

CRITICAL THINKING AND *THE STRUCTURE OF SCIENTIFIC REVOLUTIONS*

- Primary critical thinking skill: CREATIVE THINKING
- Secondary critical thinking skill: REASONING

Thomas Kuhn's *The Structure of Scientific Revolutions* can be seen, without exaggeration, as a landmark text in intellectual history.

In his analysis of shifts in scientific thinking, Kuhn questioned the prevailing view that science was an unbroken progression towards the truth. Progress was actually made, he argued, via "paradigm shifts", meaning that evidence that existing scientific models are flawed slowly accumulates—in the face, at first, of opposition and doubt—until it finally results in a crisis that forces the development of a new model. This development, in turn, produces a period of rapid change—"extraordinary science," Kuhn terms it—before an eventual return to "normal science" begins the process whereby the whole cycle eventually repeats itself.

This portrayal of science as the product of successive revolutions was the product of rigorous but imaginative critical thinking. It was at odds with science's self-image as a set of disciplines that constantly evolve and progress via the process of building on existing knowledge. Kuhn's highly creative re-imagining of that image has proved enduringly influential—and is the direct product of the author's ability to produce a novel explanation for existing evidence and to redefine issues so as to see them in new ways.

《世界思想宝库钥匙丛书》简介

《世界思想宝库钥匙丛书》致力于为一系列在各领域产生重大影响的人文社科类经典著作提供独特的学术探讨。每一本读物都不仅仅是原经典著作的内容摘要,而是介绍并深入研究原经典著作的学术渊源、主要观点和历史影响。这一丛书的目的是提供一套学习资料,以促进读者掌握批判性思维,从而更全面、深刻地去理解重要思想。

每一本读物分为3个部分:学术渊源、学术思想和学术影响,每个部分下有4个小节。这些章节旨在从各个方面研究原经典著作及其反响。

由于独特的体例,每一本读物不但易于阅读,而且另有一项优点:所有读物的编排体例相同,读者在进行某个知识层面的调查或研究时可交叉参阅多本该丛书中的相关读物,从而开启跨领域研究的路径。

为了方便阅读,每本读物最后还列出了术语表和人名表(在书中则以星号＊标记),此外还有参考文献。

《世界思想宝库钥匙丛书》与剑桥大学合作,理清了批判性思维的要点,即如何通过6种技能来进行有效思考。其中3种技能让我们能够理解问题,另3种技能让我们有能力解决问题。这6种技能合称为"批判性思维PACIER模式",它们是:

分析:了解如何建立一个观点;
评估:研究一个观点的优点和缺点;
阐释:对意义所产生的问题加以理解;
创造性思维:提出新的见解,发现新的联系;
解决问题:提出切实有效的解决办法;
理性化思维:创建有说服力的观点。

了解更多信息,请浏览www.macat.com。

THE MACAT LIBRARY

The Macat Library is a series of unique academic explorations of seminal works in the humanities and social sciences — books and papers that have had a significant and widely recognised impact on their disciplines. It has been created to serve as much more than just a summary of what lies between the covers of a great book. It illuminates and explores the influences on, ideas of, and impact of that book. Our goal is to offer a learning resource that encourages critical thinking and fosters a better, deeper understanding of important ideas.

Each publication is divided into three Sections: Influences, Ideas, and Impact. Each Section has four Modules. These explore every important facet of the work, and the responses to it.

This Section-Module structure makes a Macat Library book easy to use, but it has another important feature. Because each Macat book is written to the same format, it is possible (and encouraged!) to cross-reference multiple Macat books along the same lines of inquiry or research. This allows the reader to open up interesting interdisciplinary pathways.

To further aid your reading, lists of glossary terms and people mentioned are included at the end of this book (these are indicated by an asterisk [*] throughout) — as well as a list of works cited.

Macat has worked with the University of Cambridge to identify the elements of critical thinking and understand the ways in which six different skills combine to enable effective thinking.

Three allow us to fully understand a problem; three more give us the tools to solve it. Together, these six skills make up the PACIER model of critical thinking. They are:

ANALYSIS — understanding how an argument is built
EVALUATION — exploring the strengths and weaknesses of an argument
INTERPRETATION — understanding issues of meaning
CREATIVE THINKING — coming up with new ideas and fresh connections
PROBLEM-SOLVING — producing strong solutions
REASONING — creating strong arguments

To find out more, visit www.MACAT.com.

"《世界思想宝库钥匙丛书》提供了独一无二的跨学科学习和研究工具。它介绍那些革新了各自学科研究的经典著作,还邀请全世界一流专家和教育机构进行严谨的分析,为每位读者打开世界顶级教育的大门。"

—— 安德烈亚斯·施莱歇尔,
经济合作与发展组织教育与技能司司长

"《世界思想宝库钥匙丛书》直面大学教育的巨大挑战……他们组建了一支精干而活跃的学者队伍,来推出在研究广度上颇具新意的教学材料。"

—— 布罗尔斯教授、勋爵,剑桥大学前校长

"《世界思想宝库钥匙丛书》的愿景令人赞叹。它通过分析和阐释那些曾深刻影响人类思想以及社会、经济发展的经典文本,提供了新的学习方法。它推动批判性思维,这对于任何社会和经济体来说都是至关重要的。这就是未来的学习方法。"

—— 查尔斯·克拉克阁下,英国前教育大臣

"对于那些影响了各自领域的著作,《世界思想宝库钥匙丛书》能让人们立即了解到围绕那些著作展开的评论性言论,这让该系列图书成为在这些领域从事研究的师生们不可或缺的资源。"

—— 威廉·特朗佐教授,加利福尼亚大学圣地亚哥分校

"Macat offers an amazing first-of-its-kind tool for interdisciplinary learning and research. Its focus on works that transformed their disciplines and its rigorous approach, drawing on the world's leading experts and educational institutions, opens up a world-class education to anyone."

—— Andreas Schleicher, Director for Education and Skills, Organisation for Economic Co-operation and Development

"Macat is taking on some of the major challenges in university education ... They have drawn together a strong team of active academics who are producing teaching materials that are novel in the breadth of their approach."

—— Prof Lord Broers, former Vice-Chancellor of the University of Cambridge

"The Macat vision is exceptionally exciting. It focuses upon new modes of learning which analyse and explain seminal texts which have profoundly influenced world thinking and so social and economic development. It promotes the kind of critical thinking which is essential for any society and economy. This is the learning of the future."

—— Rt Hon Charles Clarke, former UK Secretary of State for Education

"The Macat analyses provide immediate access to the critical conversation surrounding the books that have shaped their respective discipline, which will make them an invaluable resource to all of those, students and teachers, working in the field."

—— Prof William Tronzo, University of California at San Diego

The Macat Library
世界思想宝库钥匙丛书

TITLE	中文书名	类别
An Analysis of Arjun Appadurai's *Modernity at Large: Cultural Dimensions of Globalization*	解析阿尔君·阿帕杜莱《消失的现代性：全球化的文化维度》	人类学
An Analysis of Claude Lévi-Strauss's *Structural Anthropology*	解析克劳德·列维–斯特劳斯《结构人类学》	人类学
An Analysis of Marcel Mauss's *The Gift*	解析马塞尔·莫斯《礼物》	人类学
An Analysis of Jared M. Diamond's *Guns, Germs, and Steel: The Fate of Human Societies*	解析贾雷德·M.戴蒙德《枪炮、病菌与钢铁：人类社会的命运》	人类学
An Analysis of Clifford Geertz's *The Interpretation of Cultures*	解析克利福德·格尔茨《文化的解释》	人类学
An Analysis of Philippe Ariès's *Centuries of Childhood: A Social History of Family Life*	解析菲力浦·阿利埃斯《儿童的世纪：旧制度下的儿童和家庭生活》	人类学
An Analysis of W. Chan Kim & Renée Mauborgne's *Blue Ocean Strategy*	解析金伟灿/勒妮·莫博涅《蓝海战略》	商业
An Analysis of John P. Kotter's *Leading Change*	解析约翰·P.科特《领导变革》	商业
An Analysis of Michael E. Porter's *Competitive Strategy: Techniques for Analyzing Industries and Competitors*	解析迈克尔·E.波特《竞争战略：分析产业和竞争对手的技术》	商业
An Analysis of Jean Lave & Etienne Wenger's *Situated Learning: Legitimate Peripheral Participation*	解析琼·莱夫/艾奈纳·温格《情境学习：合法的边缘性参与》	商业
An Analysis of Douglas McGregor's *The Human Side of Enterprise*	解析道格拉斯·麦格雷戈《企业的人性面》	商业
An Analysis of Milton Friedman's *Capitalism and Freedom*	解析米尔顿·弗里德曼《资本主义与自由》	商业
An Analysis of Ludwig von Mises's *The Theory of Money and Credit*	解析路德维希·冯·米塞斯《货币和信用理论》	经济学
An Analysis of Adam Smith's *The Wealth of Nations*	解析亚当·斯密《国富论》	经济学
An Analysis of Thomas Piketty's *Capital in the Twenty-First Century*	解析托马斯·皮凯蒂《21世纪资本论》	经济学
An Analysis of Nassim Nicholas Taleb's *The Black Swan: The Impact of the Highly Improbable*	解析纳西姆·尼古拉斯·塔勒布《黑天鹅：如何应对不可预知的未来》	经济学
An Analysis of Ha-Joon Chang's *Kicking Away the Ladder*	解析张夏准《富国陷阱：发达国家为何踢开梯子》	经济学
An Analysis of Thomas Robert Malthus's *An Essay on the Principle of Population*	解析托马斯·罗伯特·马尔萨斯《人口论》	经济学

An Analysis of John Maynard Keynes's *The General Theory of Employment, Interest and Money*	解析约翰·梅纳德·凯恩斯《就业、利息和货币通论》	经济学
An Analysis of Milton Friedman's *The Role of Monetary Policy*	解析米尔顿·弗里德曼《货币政策的作用》	经济学
An Analysis of Burton G. Malkiel's *A Random Walk Down Wall Street*	解析伯顿·G. 马尔基尔《漫步华尔街》	经济学
An Analysis of Friedrich A. Hayek's *The Road to Serfdom*	解析弗里德里希·A. 哈耶克《通往奴役之路》	经济学
An Analysis of Charles P. Kindleberger's *Manias, Panics, and Crashes: A History of Financial Crises*	解析查尔斯·P. 金德尔伯格《疯狂、惊恐和崩溃：金融危机史》	经济学
An Analysis of Amartya Sen's *Development as Freedom*	解析阿马蒂亚·森《以自由看待发展》	经济学
An Analysis of Rachel Carson's *Silent Spring*	解析蕾切尔·卡森《寂静的春天》	地理学
An Analysis of Charles Darwin's *On the Origin of Species: by Means of Natural Selection, or The Preservation of Favoured Races in the Struggle for Life*	解析查尔斯·达尔文《物种起源》	地理学
An Analysis of World Commission on Environment and Development's *The Brundtland Report: Our Common Future*	解析世界环境与发展委员会《布伦特兰报告：我们共同的未来》	地理学
An Analysis of James E. Lovelock's *Gaia: A New Look at Life on Earth*	解析詹姆斯·E. 拉伍洛克《盖娅：地球生命的新视野》	地理学
An Analysis of Paul Kennedy's *The Rise and Fall of the Great Powers: Economic Change and Military Conflict from 1500–2000*	解析保罗·肯尼迪《大国的兴衰：1500—2000 年的经济变革与军事冲突》	历史
An Analysis of Janet L. Abu-Lughod's *Before European Hegemony: The World System A. D. 1250–1350*	解析珍妮特·L. 阿布-卢格霍德《欧洲霸权之前：1250—1350 年的世界体系》	历史
An Analysis of Alfred W. Crosby's *The Columbian Exchange: Biological and Cultural Consequences of 1492*	解析艾尔弗雷德·W. 克罗斯比《哥伦布大交换：1492 年以后的生物影响和文化冲击》	历史
An Analysis of Tony Judt's *Postwar: A History of Europe since 1945*	解析托尼·朱特《战后欧洲史》	历史
An Analysis of Richard J. Evans's *In Defence of History*	解析理查德·J. 艾文斯《捍卫历史》	历史
An Analysis of Eric Hobsbawm's *The Age of Revolution: Europe 1789–1848*	解析艾瑞克·霍布斯鲍姆《革命的年代：欧洲 1789—1848 年》	历史

An Analysis of Roland Barthes's *Mythologies*	解析罗兰·巴特《神话学》	文学与批判理论
An Analysis of Simone de Beauvoir's *The Second Sex*	解析西蒙娜·德·波伏娃《第二性》	文学与批判理论
An Analysis of Edward W. Said's *Orientalism*	解析爱德华·W. 萨义德《东方主义》	文学与批判理论
An Analysis of Virginia Woolf's *A Room of One's Own*	解析弗吉尼亚·伍尔芙《一间自己的房间》	文学与批判理论
An Analysis of Judith Butler's *Gender Trouble*	解析朱迪斯·巴特勒《性别麻烦》	文学与批判理论
An Analysis of Ferdinand de Saussure's *Course in General Linguistics*	解析费尔迪南·德·索绪尔《普通语言学教程》	文学与批判理论
An Analysis of Susan Sontag's *On Photography*	解析苏珊·桑塔格《论摄影》	文学与批判理论
An Analysis of Walter Benjamin's *The Work of Art in the Age of Mechanical Reproduction*	解析瓦尔特·本雅明《机械复制时代的艺术作品》	文学与批判理论
An Analysis of W. E. B. Du Bois's *The Souls of Black Folk*	解析W.E.B.杜波依斯《黑人的灵魂》	文学与批判理论
An Analysis of Plato's *The Republic*	解析柏拉图《理想国》	哲学
An Analysis of Plato's *Symposium*	解析柏拉图《会饮篇》	哲学
An Analysis of Aristotle's *Metaphysics*	解析亚里士多德《形而上学》	哲学
An Analysis of Aristotle's *Nicomachean Ethics*	解析亚里士多德《尼各马可伦理学》	哲学
An Analysis of Immanuel Kant's *Critique of Pure Reason*	解析伊曼努尔·康德《纯粹理性批判》	哲学
An Analysis of Ludwig Wittgenstein's *Philosophical Investigations*	解析路德维希·维特根斯坦《哲学研究》	哲学
An Analysis of G. W. F. Hegel's *Phenomenology of Spirit*	解析G.W.F.黑格尔《精神现象学》	哲学
An Analysis of Baruch Spinoza's *Ethics*	解析巴鲁赫·斯宾诺莎《伦理学》	哲学
An Analysis of Hannah Arendt's *The Human Condition*	解析汉娜·阿伦特《人的境况》	哲学
An Analysis of G. E. M. Anscombe's *Modern Moral Philosophy*	解析G.E.M.安斯康姆《现代道德哲学》	哲学
An Analysis of David Hume's *An Enquiry Concerning Human Understanding*	解析大卫·休谟《人类理解研究》	哲学

An Analysis of Søren Kierkegaard's *Fear and Trembling*	解析索伦·克尔凯郭尔《恐惧与战栗》	哲学
An Analysis of René Descartes's *Meditations on First Philosophy*	解析勒内·笛卡尔《第一哲学沉思录》	哲学
An Analysis of Friedrich Nietzsche's *On the Genealogy of Morality*	解析弗里德里希·尼采《论道德的谱系》	哲学
An Analysis of Gilbert Ryle's *The Concept of Mind*	解析吉尔伯特·赖尔《心的概念》	哲学
An Analysis of Thomas Kuhn's *The Structure of Scientific Revolutions*	解析托马斯·库恩《科学革命的结构》	哲学
An Analysis of John Stuart Mill's *Utilitarianism*	解析约翰·斯图亚特·穆勒《功利主义》	哲学
An Analysis of Aristotle's *Politics*	解析亚里士多德《政治学》	政治学
An Analysis of Niccolò Machiavelli's *The Prince*	解析尼科洛·马基雅维利《君主论》	政治学
An Analysis of Karl Marx's *Capital*	解析卡尔·马克思《资本论》	政治学
An Analysis of Benedict Anderson's *Imagined Communities*	解析本尼迪克特·安德森《想象的共同体》	政治学
An Analysis of Samuel P. Huntington's *The Clash of Civilizations and the Remaking of World Order*	解析塞缪尔·P.亨廷顿《文明的冲突与世界秩序的重建》	政治学
An Analysis of Alexis de Tocqueville's *Democracy in America*	解析阿列克西·德·托克维尔《论美国的民主》	政治学
An Analysis of John A. Hobson's *Imperialism: A Study*	解析约翰·A.霍布森《帝国主义》	政治学
An Analysis of Thomas Paine's *Common Sense*	解析托马斯·潘恩《常识》	政治学
An Analysis of John Rawls's *A Theory of Justice*	解析约翰·罗尔斯《正义论》	政治学
An Analysis of Francis Fukuyama's *The End of History and the Last Man*	解析弗朗西斯·福山《历史的终结与最后的人》	政治学
An Analysis of John Locke's *Two Treatises of Government*	解析约翰·洛克《政府论》	政治学
An Analysis of Sun Tzu's *The Art of War*	解析孙武《孙子兵法》	政治学
An Analysis of Henry Kissinger's *World Order: Reflections on the Character of Nations and the Course of History*	解析亨利·基辛格《世界秩序》	政治学
An Analysis of Jean-Jacques Rousseau's *The Social Contract*	解析让-雅克·卢梭《社会契约论》	政治学

An Analysis of Odd Arne Westad's *The Global Cold War: Third World Interventions and the Making of Our Times*	解析文安立《全球冷战：美苏对第三世界的干涉与当代世界的形成》	政治学
An Analysis of Sigmund Freud's *The Interpretation of Dreams*	解析西格蒙德·弗洛伊德《梦的解析》	心理学
An Analysis of William James' *The Principles of Psychology*	解析威廉·詹姆斯《心理学原理》	心理学
An Analysis of Philip Zimbardo's *The Lucifer Effect*	解析菲利普·津巴多《路西法效应》	心理学
An Analysis of Leon Festinger's *A Theory of Cognitive Dissonance*	解析利昂·费斯汀格《认知失调论》	心理学
An Analysis of Richard H. Thaler & Cass R. Sunstein's *Nudge: Improving Decisions about Health, Wealth, and Happiness*	解析理查德·H.泰勒/卡斯·R.桑斯坦《助推：如何做出有关健康、财富和幸福的更优决策》	心理学
An Analysis of Gordon Allport's *The Nature of Prejudice*	解析高尔登·奥尔波特《偏见的本质》	心理学
An Analysis of Steven Pinker's *The Better Angels of Our Nature: Why Violence Has Declined*	解析斯蒂芬·平克《人性中的善良天使：暴力为什么会减少》	心理学
An Analysis of Stanley Milgram's *Obedience to Authority*	解析斯坦利·米尔格拉姆《对权威的服从》	心理学
An Analysis of Betty Friedan's *The Feminine Mystique*	解析贝蒂·弗里丹《女性的奥秘》	心理学
An Analysis of David Riesman's *The Lonely Crowd: A Study of the Changing American Character*	解析大卫·理斯曼《孤独的人群：美国人社会性格演变之研究》	社会学
An Analysis of Franz Boas's *Race, Language and Culture*	解析弗朗兹·博厄斯《种族、语言与文化》	社会学
An Analysis of Pierre Bourdieu's *Outline of a Theory of Practice*	解析皮埃尔·布尔迪厄《实践理论大纲》	社会学
An Analysis of Max Weber's *The Protestant Ethic and the Spirit of Capitalism*	解析马克斯·韦伯《新教伦理与资本主义精神》	社会学
An Analysis of Jane Jacobs's *The Death and Life of Great American Cities*	解析简·雅各布斯《美国大城市的死与生》	社会学
An Analysis of C. Wright Mills's *The Sociological Imagination*	解析C.赖特·米尔斯《社会学的想象力》	社会学
An Analysis of Robert E. Lucas Jr.'s *Why Doesn't Capital Flow from Rich to Poor Countries?*	解析小罗伯特·E.卢卡斯《为何资本不从富国流向穷国？》	社会学

An Analysis of Émile Durkheim's *On Suicide*	解析埃米尔·迪尔凯姆《自杀论》	社会学
An Analysis of Eric Hoffer's *The True Believer: Thoughts on the Nature of Mass Movements*	解析埃里克·霍弗《狂热分子：群众运动圣经》	社会学
An Analysis of Jared M. Diamond's *Collapse: How Societies Choose to Fail or Survive*	解析贾雷德·M.戴蒙德《大崩溃：社会如何选择兴亡》	社会学
An Analysis of Michel Foucault's *The History of Sexuality Vol. 1: The Will to Knowledge*	解析米歇尔·福柯《性史（第一卷）：求知意志》	社会学
An Analysis of Michel Foucault's *Discipline and Punish*	解析米歇尔·福柯《规训与惩罚》	社会学
An Analysis of Richard Dawkins's *The Selfish Gene*	解析理查德·道金斯《自私的基因》	社会学
An Analysis of Antonio Gramsci's *Prison Notebooks*	解析安东尼奥·葛兰西《狱中札记》	社会学
An Analysis of Augustine's *Confessions*	解析奥古斯丁《忏悔录》	神学
An Analysis of C. S. Lewis's *The Abolition of Man*	解析C.S.路易斯《人之废》	神学

图书在版编目（CIP）数据

解析托马斯·库恩《科学革命的结构》：汉、英 / 乔·赫德桑, 约瑟夫·滕德勒著. 丰国欣译. -- 上海：上海外语教育出版社, 2020 (2024重印)
（世界思想宝库钥匙丛书）
ISBN 978-7-5446-6500-1

Ⅰ. ①解⋯ Ⅱ. ①乔⋯ ②约⋯ ③丰⋯ Ⅲ. ①科学哲学—汉、英 Ⅳ. ①N02

中国版本图书馆CIP数据核字（2020）第143391号

This Chinese-English bilingual edition of *An Analysis of Thomas Kuhn's The Structure of Scientific Revolutions* is published by arrangement with Macat International Limited. Licensed for sale throughout the world.

本书汉英双语版由Macat国际有限公司授权上海外语教育出版社有限公司出版。供在全世界范围内发行、销售。

图字：09 – 2018 – 549

出版发行：**上海外语教育出版社**
（上海外国语大学内）　邮编：200083
电　　话：021-65425300（总机）
电子邮箱：bookinfo@sflep.com.cn
网　　址：http://www.sflep.com
责任编辑：李振荣

印　　刷：上海新华印刷有限公司
开　　本：890×1240　1/32　印张 7.625　字数 156千字
版　　次：2020 年 9 月第 1 版　2024 年 1 月第 2 次印刷

书　　号：ISBN 978-7-5446-6500-1
定　　价：30.00 元

本版图书如有印装质量问题，可向本社调换
质量服务热线：4008-213-263